Let's Do This.

First things first—thank you. Whether you bought this book, got it as a gift, "borrowed" it, or just happened to stumble across it, you're here for a reason. And that reason is bigger than just a test score—it's about unlocking opportunities for your future.

You now hold in your hands the **Enhanced ACT Manual**, a tool designed to give you an edge. Inside, you'll find strategies, tips, and practice that can help you work smarter, think sharper, and score higher. But here's the truth: this book can't do the work for you. Your results will ultimately depend on your focus, your effort, and your consistency.

I've spent years traveling across the continental U.S., helping over 10,000 students, teachers, faculty, and staff prepare for the biggest tests of their lives—ACT, SAT, Digital SAT, TSIA, TSIA2, LEAP, STAAR, NC1 Math, WorkKeys, the Pharmacy Technician Exam (PTCE), the ExCPT, the CMA Exam, and more. I've seen firsthand what happens when someone takes the right tools and puts in the work—they win.

So, here's the deal: read it, study it, apply it. Commit it to memory. Use it until the strategies become second nature. If you do, this book won't just help you take the ACT—it'll help you beat it.

Your future is calling. Let's get started.

LET'S DO THIS.

READ IT. STUDY IT. APPLY IT.

Commit it to memory. Use it until the strategies become second nature.

YOUR FUTURE IS CALLING. LET'S GET STARTED.

Table of Contents

Chapter 1: Notable Changes

One of the biggest challenges students face when taking the ACT has always been **time**—feeling rushed, pressured, and barely finishing each section. The creators of the ACT have listened to feedback from across the country and made big changes to help.

The Enhanced ACT gives you more time to think through each question by shortening the test overall, reducing the number of questions, and providing shorter reading passages (goodbye, sleepy reading section!). You'll still need focus and strategy, but now you'll have a calmer, less frantic testing experience from start to finish.

Another major change: the Science section is now optional. If you feel strong in science, go for it—it could boost your score. But if you know you'll be mentally drained by the end, you can skip it without penalty.

These changes are designed to give you more control, reduce pressure, and help you show what you know on test day. Let's take a closer look at how the new Enhanced ACT compares to the classic version.

Classic_vs_Enhanced_ACT_Comparison

Feature	Classic ACT	Enhanced ACT
Total Test Time	2 hours 55 minutes (+ optional 40 minutes for Writing)	Shorter overall test time
Number of Questions	215 questions	Fewer questions
Average Time per Question	Shorter time per question	More time per question
Reading Section Passages	4 long passages	Shorter passages
Science Section	Required	Optional

Section	Classic	Enhanced	Classic	Enhanced	Extra time per question
English	75 questions	50 questions	45 mins	35 mins	6 seconds
Math	60 questions (5 choices)	45 questions (4 choices)	60 mins	50 mins	7 seconds
Reading	40 questions	36 questions	35 mins	40 mins	14 seconds
Science	40 questions	40 questions	35 mins	40 mins	7 seconds
Writing	1	1	40	40	N/A

Six to fourteen seconds might not sound like much, but on the ACT, every second is gold. Those extra moments stack up, giving you the breathing room you need to think clearly and work strategically. But here's the truth—time alone won't save you. If you don't know the material, the clock can't help you. On the other hand, if you've mastered the content, you'll move through the easy questions quickly and have extra time to tackle the tough ones. That's the game: speed and accuracy working together.

So, here's the plan—let's sharpen your skills. We'll start with English and finish strong with Science. But first, a little trivia to loosen things up: Why is the ACT in the order it's in?

(Answer: It's alphabetical. No hidden meaning, just the ABCs at work.)

Now, let's get serious. You've heard "practice makes perfect," but the real truth is—practice makes permanent. How you train is how you'll perform. If you approach this with focus and discipline, you'll get the most out of every session, so treat every practice like the real thing. No phones. No tablets. No music. No movies. Especially when you're taking a practice test. This is your time to lock in, push yourself, and prepare to crush the Enhanced ACT.

ENGLISH

Chapter 2: English Stretch

“Prepare, prepare, prepare. When the pressure is on, you don’t rise to the occasion. You fall to your highest level of preparation.”
Chris Voss

“Practice isn't the thing you do once you're good. It's the thing you do that makes you good.”
Malcolm Gladwell

"Great things come from hard work and perseverance. No excuses."
Kobe Bryant

Game time isn’t when you start learning the plays — it’s when you show off the work you’ve already put in. When I played organized sports, every practice started the same way — with stretching, warm-ups, drills, or a scrimmage. You don’t just run onto the field or court cold. Even the pros stick to this routine. In the same way, we’re starting with fundamental content warm-ups to sharpen your skills, then finishing with a full-on, timed “scrimmage” to simulate the real test. This way, when it’s game time, you’re already warmed up, focused, and ready to win.

Question 1: Choose the best version: The dog chased it's tail around the yard.

A. No Change

B. its

C. its'

D. it

Question 2: Which choice maintains parallel structure: She likes dancing, to swim, and biking.

A. dancing, swimming, and biking.

B. dance, to swim, and bike.

C. dancing, to swim, and to bike.

D. No change

Question 3: Choose the most concise version: Due to the fact that it was raining, the game was canceled.

A. Because it was raining

B. Owing to the rain

C. Since it started raining out the blue

D. No change

Question 4: Select the grammatically correct sentence.

A. Neither the players or the coach was ready.

B. Neither the players nor the coach was ready.

C. Neither the players nor the coach were ready.

D. Neither the player or the coach were ready.

Question 5: Choose the best punctuation: <u>My brother who lives in Houston</u> is visiting.

A. brother, who lives in Houston,

B. brother who lives in Houston,

C. brother, who lives in Houston

D. No change

Question 6: Choose the best revision: Maria sings beautifully, she plays the piano well too.

A. Maria sings beautifully she plays the piano well too.

B. Maria sings beautifully and she plays the piano well too.

C. Maria sings beautifully; she plays the piano well too.

D. No change

Question 7: Select the sentence with correct subject-verb agreement.

A. Each of the players have a jersey.

B. Each of the players are ready.

C. Each of the players were asked to arrive early.

D. Each of the players has a jersey.

Question 8: Choose the most logical transition: He loves classical music. <u>Nevertheless</u>, he attends every rock concert.

A. Therefore

B. No change

C. For example

D. As a result

Question 9: Identify the best placement for the modifying phrase: Barking loudly, the mailman was startled by the dog.

A. The dog startled the mailman barking loudly.

B. The mailman was startled barking loudly by the dog.

C. Barking loudly, the dog startled the mailman.

D. The mailman was startled by the dog barking loudly.

Question 10: Select the clearest and most effective version: The athlete who was injured, she could not play.

A. She could not play, the athlete who was injured.

B. She, the athlete who was injured, could not play anymore.

C. The injured athlete could not play.

D. No change

Question 11: Choose the best word choice: The desert air was so dry, it <u>literally</u> sucked the moisture from your skin.

A. really

B. practically

C. absolutely

D. No change

Question 12: Pick the most appropriate verb tense: By next summer, she ____ graduated.

A. has

B. had

C. will have

D. would have

Question 13: Choose the correct punctuation: My favorite meal includes steak, potatoes and green beans, and apple pie.

A. steak, potatoes, green beans, and apple pie

B. steak potatoes green beans and apple pie

C. steak, and potatoes with green beans, and apple pie

D. No Change

Question 14: Which version avoids wordiness? At this point in time, we are currently considering the options.

A. At this current time, we are thinking.

B. We are considering the options.

C. No change

D. We are presently and currently reviewing options.

Question 15: Choose the clearest sentence revision: It was the book, the one with the torn pages, that the teacher assigned.

A. The teacher assigned it, the book, the one with torn pages.

B. The book was the one that the teacher, who had chosen it, assigned.

C. The teacher assigned the book, which had torn pages, it was that one.

D. The teacher assigned the book with the torn pages.

Question 16: Which version maintains proper verb tense?

A. Yesterday, she walked to school and met her friend.

B. Yesterday, she walks to school and met her friend.

C. Yesterday, she walking to school and meets her friend.

D. Yesterday, she walk to school and meet her friend.

Question 17: Choose the clearest and most concise version:

At this point in time, we are currently reviewing the application.

A. We are reviewing the application.

B. At this point, we are currently reviewing the application.

C. Currently, we are reviewing the application at this point in time.

D. We are currently reviewing the application at this time.

Question 18: Select the sentence with correct subject-verb agreement.

A. The team of players are celebrating their victory.

B. The team of players is celebrating their victory.

C. The teams of players is celebrating their victory.

D. The teams of player is celebrating their victory.

Question 19: Choose the sentence with correct punctuation:

A. The books, which are old, are on the top shelf.

B. The books which are old, are on the top shelf.

C. The books which are old are on, the top shelf.

D. The books which, are old are on the top shelf.

Question 20: Which is the best revision for clarity?

The reason she left is because she was tired.

A. She left because she was tired.

B. The reason why she left is because she was tired.

C. Because she was tired that is the reason she left.

D. She was tired, that's the reason she left.

Question 21: Select the correct sentence.

A. Running through the woods, the trees blocked his path.

B. Running through the woods, he saw the trees blocking his path.

C. He saw the trees running through the woods, blocking his path.

D. Blocking his path, he ran through the trees in the woods.

Question 22: Choose the most effective transition word:

She was tired from the hike. _______, she decided to keep walking.

A. Nevertheless

B. Similarly

C. For example

D. Meanwhile

Question 23: Choose the clearest and most concise sentence.

A. It is often the case that people do not get enough sleep.

B. Often, people do not get enough sleep.

C. It is often true that people are not sleeping enough.

D. There is a tendency for people to sleep less.

Question 24: Choose the sentence with correct punctuation.

A. She packed a tent, a flashlight and, a sleeping bag.

B. She packed a tent a flashlight, and a sleeping bag.

C. She packed a tent, a flashlight, and a sleeping bag.

D. She packed a tent a flashlight and a sleeping bag.

Question 25: Choose the best version:

The committee made their decision yesterday.

A. No change

B. The committee made its decision yesterday.

C. The committee, they made a decision yesterday.

D. The committee had made their decision yesterday.

Question 26: Choose the grammatically correct sentence.

A. Him and I went to the movies.

B. He and me went to the movies.

C. He and I went to the movies.

D. Me and him went to the movies.

Question 27: Identify the sentence with proper parallel structure.

A. The job requires attention to detail, being on time, and that you are organized.

B. The job requires attention to detail, punctuality, and being organized.

C. The job requires that you pay attention to detail, that you're on time, and being organized.

D. The job requires attention to detail, punctuality, and organization.

Question 28: Which version is most appropriate in tone for a formal essay?

A. The experiment was kinda a bust.

B. The experiment didn't really work out.

C. The experiment was unsuccessful.

D. The experiment was a total fail.

Question 29: Choose the best revision: If I was tall, I could reach the top shelf.

A. If I were taller, I could reach the top shelf.

B. No Change

C. If I had been taller, I can reach the top shelf.

D. If I were tall, I reach the top shelf.

Question 30: Choose the sentence with correct apostrophe usage.

A. The dogs bone was buried in the yard.

B. The dog's bone was buried in the yard.

C. The dogs' bone was buried in the yard.

D. The dog bone's was buried in the yard.

Question 31: Select the clearest revision:

He was late because of the fact that he missed the bus.

A. He was late because he missed the bus.

B. Because of the bus being missed, he was late.

C. He missed the bus and thus he was late.

D. Due to the missing of the bus, he became late.

Question 32: Choose the most appropriate verb tense:

By the time we arrived, the concert ______.

A. has started

B. had started

C. will have started

D. starts

Question 33: Identify the correct sentence.

A. She loves to read, she hates writing.

B. She loves to read: she hates writing.

C. She loves to read, but she hates writing.

D. She loves to read and she hates writing.

Question 34: Choose the correct usage:

Everyone should bring ______ own laptop.

A. his or her

B. their

C. his and her

D. one's

Question 35: Choose the sentence with proper subject-verb agreement.

A. Neither the players nor the coach are happy.

B. Neither the coach nor the players is happy.

C. Neither the coach nor the players aren’t happy.

D. Neither the players nor the coach is happy.

Question 36: Choose the best sentence: The class was not only interesting but also it was informative.

A. No change

B. The class not only was interesting but also informative.

C. The class was not only interesting but also informative.

D. Not only was the class interesting, it was informative.

Question 37: Select the sentence with correct punctuation.

A. After dinner we had dessert.

B. After dinner, we had dessert.

C. We had, dessert after dinner.

D. We, had dessert after dinner.

Question 38: Which sentence uses modifiers correctly?

A. Covered in syrup, he ate the pancakes.

B. He ate the pancakes covered in syrup.

C. He covered in syrup, ate the pancakes.

D. Covered in syrup, the pancakes were eaten by him.

Question 39: Choose the best version: The reason why she left early is because she was feeling sick.

A. No change

B. The reason she left early is because she was feeling sick.

C. She left early because she was feeling sick.

D. The reason she was feeling sick is why she left early.

Question 40: Choose the clearest sentence.

A. Due to the fact that it was raining, we canceled the game.

B. Because it was raining, we canceled the game.

C. It was raining, therefore, we canceled the game.

D. Raining it was, so we canceled the game.

Question 41: Which sentence uses colons correctly?

A. She bought: apples, oranges, and grapes.

B. She bought the following: apples, oranges, and grapes.

C. She bought the following apples, oranges: and grapes.

D. She bought the following apples: oranges and grapes.

Question 42: Select the correct pronoun usage.

A. Her and I went to the store.

B. She and me went to the store.

C. She and I went to the store.

D. Me and her went to the store.

Question 43: Choose the sentence that uses commas correctly.

A. My dad who is a doctor, works at the hospital.

B. My dad, who is a doctor works at the hospital.

C. My dad who is a doctor works at the hospital.

D. My dad, who is a doctor, works at the hospital.

Question 44: Choose the correct word to complete the sentence:

You're going to need more time to ______ your project.

A. complete

B. complement

C. compete

D. compliment

Question 45: Which version is most grammatically correct?

A. The team have won their last five games.

B. The team has won its last five games.

C. The team has won their last five games.

D. The team have won its last five games.

Question 46: Choose the sentence that avoids redundancy.

A. The end result was a complete success.

B. The result was a complete success.

C. The final outcome result was a complete success.

D. The result that occurred in the end was a complete success.

Question 47: Choose the sentence that correctly uses parallel structure.

A. He likes singing, to run, and hunting.

B. He likes to sing, running, and hunt.

C. He likes singing, running, and hunting.

D. He likes to sing, to run, and hunting.

Question 48: Identify the sentence with correct subject-verb agreement.

A. The list of items are on the table.

B. The list of items is on the table.

C. The lists of item is on the table.

D. The items of list are on the table.

Question 49: Choose the correct sentence.

A. Neither my mom or my dad like spicy food.

B. Neither my mom nor my dad likes spicy food.

C. Neither my mom nor my dad like spicy food.

D. Neither my mom or my dad likes spicy food.

Question 50: Choose the best closing sentence for a persuasive paragraph.

A. This is my opinion.

B. In conclusion, thank you for reading.

C. That's why I believe recycling is important for our planet.

D. Recycling is a thing we can do.

Chapter 3: English Warm-up and Drills

As we move into the warm-up phase of this study book, it's time to check out our ACT *Scouting Report.* Just like in sports, a scouting report is all about preparation. You study your opponent's tendencies, their favorite plays, their strengths, and their weaknesses—so when it's game time, you're ready.

The ACT isn't a mystery once you know how it's designed. The order of the sections, the types of questions, the traps they like to set—it's all part of the game plan. And if you know the playbook ahead of time, you've already got the edge.
Think about it: no team walks into a championship blind. They watch film, break down strategy, and study the details. That's exactly what we're doing here: knowing the ACT inside and out so nothing catches us off guard.

Now, let's open the scouting report, see exactly what we're up against, and continue preparing.

SCOUTING
REPORT
KNOW THE
PLAY.
MASTER THE GAME.
MOORE SUBSTANCE
TEST PREP

Most of the English portion is on Grammar. When was the last time you took an actual grammar class? Grammar is typically taught in grade school, so it's probably been a while. Here are some things (grammar usage and composition/rhetorical) to focus on and remember for your test.

Essential Grammar Rules with Examples

Subject-Verb Agreement:
The subject and verb must agree in number.
Example: The dog barks. (Not: The dog bark.)

Verb Tense Consistency:
Keep verb tense consistent within a sentence or related sentences.
Example: Sarah was running when she tripped.

Pronoun Agreement:
Pronouns must agree with their antecedents.
Example: Everyone must bring his or her ID.

Comma Usage:
Use commas to separate items in a list or after introductory elements.
Example: After the meeting, we went to lunch.

Apostrophes:
Use apostrophes to show possession or contractions.
Example: It's time to go. (It is)

Modifiers: Place modifiers near the word they modify to avoid confusion.

Example: Walking to school, she saw a deer.

Rhetorical Skills Strategies with Examples

Adding/Deleting Information:

Read the surrounding context before deciding to add/delete.

Example: If the sentence is redundant, choose to delete.

Wordiness:

Choose the most concise, clear option. ACT frowns upon redundancy and unnecessary words. ALWAYS keep it simple without leaving any relevant information out.

Example: Instead of 'in the event that', use 'if'.

Relevance:

Eliminate off-topic or irrelevant info.

Example: If a sentence doesn't support the paragraph's focus, remove it.

Transition Words:

Use transition words that accurately reflect the relationship between ideas.

Example: Use 'however' to show contrast.

KEY Test-Taking Tips

Answer every question—no penalty for guessing (Shoot your shot. Leaving a question blank is an automatic failure. If you at least randomly guessed, you have a 25% chance, which is better than ZERO.

Read the full sentence when answering a question: beware of **NOT** or **EXCEPT**.

Use the process of elimination. You may not always know the right answer, but most students can pick at least one or two answers that violate a grammar rule or that sound wrong. Sometimes you may not understand why the last remaining answer is correct, but recognizing and eliminating the three you know are wrong will leave you with the right answer, even if you don't understand why or cannot explain how it's right. Always eliminate incorrect answers first whenever the right choice isn't obvious.

Beware of answer choices that are too wordy or overly complicated. The ACT prefers **concise answers.**

Reread the sentences before and after what's being asked quickly if stuck—**context and consistency are everything.**

Two similar English answers are both wrong. They won't ask you to choose between the better options. Example: (A) Todd went to the store. He bought a new pair of shoes. (B) Todd went to the store; he bought a new pair of shoes. Both of those would be wrong because they're interchangeable. Another example: (B) In the same way (C) Similarly (D) Likewise. All three of those answer choices would be wrong because they are all transition words that connect the sentence in the same way.

Always choose the answer that makes the sentence the clearest (easiest for everyone to understand), for example: (A) They (B) Them (C) Both kids, or (D) Sally and Sam. Which answer choice makes it the clearest exactly who we're talking about? (D) Sally and Sam.

Common ACT Traps to Scout Out

Who vs. Whom

- Quick Trick: If you can replace it with he/she, use who. If you can replace it with him/her, use whom.
    - Example: Who/Whom is calling? → He is calling → ✅ Who.
    - To who/whom should I speak? → To him → ✅ Whom.

Affect vs. Effect

- Quick Trick: Affect is usually a verb (an action), while effect is usually a noun (a result).
    - Example: The storm will affect travel plans. (verb)
    - The storm had a big effect on travel plans. (noun)
- Memory Hook: Affect = Action. Effect = End result.

Its vs. It's

- Quick Trick: It's = It is (always a contraction). Its = possessive (belongs to it).
    - Example: It's raining outside. → It is raining outside.
    - The dog wagged its tail. (possessive)

Than vs. Then

- Quick Trick: Than compares, then tells time/order.
    - Example: She is taller than her brother.
    - We ate dinner, then went to the movie.

Their / They're / There

- Quick Trick:
    - Their = ownership (their books).
    - They're = they are (They're going home).
    - There = a place (over there).

Which vs. That

- Quick Trick: Use **which** (with a comma) when the clause is *extra information.* Use **that** (no comma) when the clause is *essential to the meaning.*
    - Example: The park installed new benches, **which** were painted bright blue. (extra info → nonrestrictive)
    - Example: The benches **that** were painted bright blue are reserved for the art fair. (essential info → restrictive)
- Memory Hook: If the sentence still makes sense without the clause → use **which** with a comma. If removing the clause changes the meaning → use **that** with no comma.

Time management is crucial to success and helps limit stress on a timed test. Don't spend more than 35 seconds per question on English. Give every question an equal chance because they are all worth the same number of points, and English isn't designed to get progressively more challenging. The last passage might be the easiest one for you, but if you use all your time trying to be perfect or overthinking the early questions, you'll run out of time for the remaining ones and be left picking anything to avoid leaving questions blank. If you use no more than 35 seconds per question (50 total), you'll have almost six minutes remaining to double-check answers you may have been unsure about.

Strategies are only as good as the practice behind them. To sharpen your skills, let's move into some ACT passage-style "drills."

Passage 1

The student council has been discussing ways to improve school spirit. One suggestion was to host a monthly theme day, where students could dress according to fun and creative themes. (1) However, there is concerns among teachers about potential distractions. The principal, who supports student engagement, has asked for a detailed proposal.

The council members decided to (2) comprise of representatives from each grade level, in order to ensure everyone's voice is heard. After gathering suggestions, they agreed that (3) having theme days not only increase morale but also builds a stronger sense of community. The proposal will (4) likely be submitting to the administration by the end of the month.

1. A. NO CHANGE
   B. there are concerns
   C. their is concerns
   D. there is concern

2. A. NO CHANGE
   B. be comprised of
   C. consist of
   D. compose of

3. A. NO CHANGE
   B. not only increases morale but also builds

C. not only increase morale but also build

D. increases not only morale but also builds

4. A. NO CHANGE

   B. likely submit

   C. likely be submitted

   D. likely have submitted

Passage 2

Every summer, the local community theater hosts a play in the park. (1) It's an event who attracts people of all ages. Families bring picnic blankets, children chase fireflies, and actors perform under the open sky. The director (2) said the tradition, which started ten years ago, has grown into a beloved summer ritual. Many residents say that (3) it bring them a sense of unity and joy, reminding them of the importance of art in everyday life. The theater group (4) were preparing for this year's performance of 'A Midsummer Night's Dream.'

1. A. NO CHANGE

   B. whom attracts

   C. that attracts

   D. which attract

2. A. NO CHANGE

   B. says the tradition

   C. had said the tradition

   D. saying the tradition

3. A. NO CHANGE
   B. it brings
   C. they bring
   D. bringing

4. A. NO CHANGE
   B. was preparing
   C. is preparing
   D. has been prepare

Passage 3

During the science fair, Maya presented her project on renewable energy. (1) Her display included charts, diagrams, and a small model wind turbine she had constructed by herself. Many attendees were impressed not only by her technical skills (2) but also she's ability to explain complex concepts in a simple way. Maya had spent weeks preparing for the event, and it was evident in the details of her presentation. One judge commented, "her project (3) demonstrate both creativity and scientific understanding." By the end of the evening, Maya (4) have been awarded first place in her category.

1. A. NO CHANGE
   B. she herself had constructed
   C. that she constructed it
   D. she constructed herself

2. A. NO CHANGE
   B. but also her ability

C. but also she had the ability

D. and also her ability

3. A. NO CHANGE

   B. demonstrates

   C. is demonstrating

   D. demonstrated

4. A. NO CHANGE

   B. was awarded

   C. had been awarded

   D. has been awarded

Passage 4

Lena had always enjoyed baking, but she never considered (1) turning her hobby into a business. That changed when her friends started placing regular orders for cookies and cakes. (2) They told her she should of started a small bakery, and Lena, after much thought, decided to give it a try. She researched local permits, experimented with recipes, and created a business plan. One of her first challenges (3) were finding a suitable location, but she managed to rent a small storefront near downtown. (4) Lena's Bakery quickly became popular, thanks to word-of-mouth and social media buzz.

1. A. NO CHANGE

   B. turning her hobby to a business

C. turning her hobby as a business

D. to turn her hobby into a business

2. A. NO CHANGE

B. should have started

C. should had started

D. should start

3. A. NO CHANGE

B. was

C. are

D. is

4. A. NO CHANGE

B. Lena's bakery is quickly becoming popular fast

C. Lena's bakery were quick to become popular

D. Lena's Bakery become popular quickly

Passage 5

(1) Carlos always dreamed of traveling the world. After graduating from college, he saved money diligently and finally booked a ticket to Europe. (2) He planned a detailed itinerary, including visits to historic cities, eating local cuisines, and attends language classes. (3) Although he had a few setbacks, like missed trains and loosed luggage, he maintained a positive attitude. (4) The trip taught him independence, adaptability, and how to appreciates different cultures.

1. A. NO CHANGE

B. dreamed to travel

C. dreams of traveling

D. was dreaming of travel

2. A. NO CHANGE

   B. eating local cuisines, and attending language classes

   C. eats local cuisines, and attending language classes

   D. eating local cuisines, and to attend language classes

3. A. NO CHANGE

   B. lose luggage

   C. lost luggage

   D. loose luggage

4. A. NO CHANGE

   B. how to appreciate

   C. how he appreciates

   D. that he appreciated

Chapter 4: English Scrimmage

You've stretched. You've warmed up. You've run the drills. Now it's time to scrimmage. This is where practice meets pressure. "You're about to take a practice (20 questions) ACT English test—crafted to mirror the style, pacing, and challenge of the real exam."

Remember—scrimmages aren't about perfection. They're about learning how to play under game-time conditions. Every mistake here is just film for us to study, and every win helps build momentum, so lock in, play smart, and let's see how you perform when the whistle blows.

If you're taking the full practice test, then set your timer for 14 minutes. If you want to take one passage at a time (10 questions), then set your timer for seven minutes and then review after each one. Both of those paces will have you ready (time management) for test day. Trust your training, use your strategies, and don't let the clock rattle you.

Make note of any type of questions you want to build on, so you can practice those online or with a teacher/tutor later. It's more than one way to prepare for and reach your ACT score goals.

The Solar Bench Project

[1]

In 2018, a group of high school students in Raleigh, North Carolina, came up with an **unusual** 1 idea: build a park bench that could charge phones using solar power. [A] The students wanted to combine environmental responsibility with something practical.

[2]

At first, the group's plans were modest. [B] They hoped to place a single bench outside the school library, but soon they realized that the idea might appeal to the wider community. [C] The bench's wooden frame **is** 2 designed by students in shop class, while the solar panels were obtained through a small grant from a local energy company.

[3]

The students encountered several setbacks. [D] For example, the wiring system overheated during the first trial run. The students were initially disappointed**, however** 4 they worked with a local electrician to correct the problem.

[4]

Subsequently, 5 the bench was finally installed in the spring. [E] A ribbon-cutting ceremony was held**, that** 6 **pulled** 7 the attention of local news outlets. **The students, proud of their accomplishment,** 8 now had to consider a new challenge: how to expand the project **beyond one bench.** 9

1. A. No change

B. weird

C. strange

D. oddball

2. F. No change

F. were

H. was

J. wasn't

3. The author is considering deleting the first sentence at point [C], "The bench's wooden frame…". Should it be kept or deleted?

A. Kept, because it provides specific details that show how the project was carried out.

B. Kept, because it explains how innovative they were.

C. Deleted, because it distracts from the focus on challenges.

D. Deleted, because it repeats unnecessary information about the park bench.

4. F. No change

G. , but however,

H. : however

J. ; however,

5. A. No change

B. Despite these difficulties,

C. On the other hand,

D. Additionally,

6. F. No change

G. it

H. which

J. , which

7. A. No change

B. lured

C. gathered

D. drew

8. Which choice would NOT be an acceptable alternative to the underlined portion?

F. The students –proud of their accomplishment–

G. The students (proud of their accomplishment)

H. The students "proud of their accomplishment"

J. Delete the commas and the phrase between them.

9. The phrase "beyond one bench" could best be replaced with which of the following to clarify the students' future goals?

A. in every single library across the globe

B. by constructing additional solar benches elsewhere

C. because one bench was clearly not enough

D. by tearing down the first bench and rebuilding it

10. The writer is considering adding the following sentence to the essay: "The project gave students a sense of how creative ideas can move from sketches to real-world results." Where would this sentence most logically be placed?

F. Point [1]

G. Point [2]

H. Point [3]

J. Point [5]

Community Mural

[1]

After the success of the solar bench project, the same group of students considered another idea;[1] painting a mural on the blank brick wall of the community center. [A] The students wanted the mural to celebrate the town's history while also adding color to the busy street.

[2]

At first, some residents were doubtful. One person said, "Murals are way too messy for me, and the paint always fades."[12] Another argued that the wall should be left blank, believing that simplicity was best. [B] However, the students scheduled a meeting at the community center to present their design sketches, which depicted scenes of local parks, factories, and festivals.

[3]

The sketches persuaded many people to support the project. [C] The students raised money by hosting a pancake breakfast. They also applied for a grant from the town's cultural arts fund.[14] Together, these efforts supplied enough money to buy paint, brushes, and ladders. They were elated to get started.

[4]

Clearly,[15] once the work began, the students realized that creating a mural was much more physically demanding than painting an indoor picture. They needed scaffolding to reach the top of the wall, and they spent hours using paint to get it ready and to make the colors correct.[16] [D] Volunteers from the neighborhood brought snacks and encouraged the painters to keep going.[17]

[5]

After three weeks of hard work, the mural was finished. The wall now showed a vivid timeline of the town, beginning with the founding of the local mill and ending with

children playing in a modern playground. [E] On the day of the unveiling, music played, families gathered, and people took <u>pictures</u> in front of the mural.
18

11. A. No change

B. idea

C. idea.

D. idea:

12. Which of the following is the best alternative for the underlined quote?

F. Murals are a messy thing, and the paint is going to fade away.

G. Murals are messy, and the paint will fade.

H. Murals are untidy, and that paint will surely eventually fade out.

J. Murals being messy and the paint fading was the issue.

13. At Point [B], the writer is considering adding the sentence: Some of the students were nervous to present their work, worried the adults wouldn't take them seriously. Should the writer add this sentence here?

A. Yes, because it highlights the students' perspective before the meeting.

B. Yes, because it explains why the sketches included local parks and festivals.

C. No, because it shifts the focus away from the reactions of the residents.

D. No, because it introduces information that is repeated later.

14. Which choice effectively combines the two sentences?

F. A pancake breakfast was hosted by the students to raise money and they also applied for a grant from the town's cultural arts fund.

G. Hosting a pancake breakfast, the students raised money, and also they applied for a grant from the town's cultural arts fund.

H. The students raised money by hosting a pancake breakfast and applied for a grant from the town's cultural arts fund.

J. A pancake breakfast was hosted by the students to raise money, they also applied for a grant from the town's cultural arts fund.

15. A. No Change

B. Similarly,

C. Indeed,

D. However,

16. Given that all the statements are true, which choice offers the most precise description?

F. No change

G. They used a lot of time mixing paint until it looked right.

H. They spent hours mixing paint so that the colors matched their designs.

J. They mixed paint for a long time to make the shades they wanted.

17. Which choice would best emphasize the community's involvement in the mural project?

A. No change

B. Paintings like murals often need approval from a city committee.

C. Some students thought about entering their mural in a national competition.

D. Murals are often painted in many cities across the country with community help.

18. F. No change

G. pictures,

H. pictures:

J. pictures–

19. At Point [E], the writer is considering adding, "Some longtime residents became emotional, saying the mural reminded them of their childhood." Should the writer add this sentence here?

A. Yes, because it shows how the mural affected community members.

B. Yes, because it repeats details already in the paragraph.

C. No, because it does not provide specific details about the unveiling.

D. No, because it distracts from the description of the students' work.

20. The writer wants to add the following sentence to the essay: "With community support growing, the students turned their attention to funding the mural."

Where would this sentence most logically be placed?

F. Point [A]

G. Point [B]

H. Point [C]

J. The sentence should **not** be added because the paragraph already contains sufficient detail.

Congratulations—you just powered through the entire English section scrimmage! That's no small feat. You battled through grammar traps, style questions, and rhetorical shifts while keeping the clock in check.

Here's what this means for you:

- **You've built stamina.** Endurance matters on test day, and you just proved you can handle a full section.
- **You've tested your strategies.** From spotting subject-verb agreement to trimming wordiness, you've seen where your game plan works—and where you might need to tighten it up.
- **You've learned under pressure.** The ACT is as much about pacing as it is about knowledge. Practicing with timing makes you sharper and more composed when the stakes are real.

Take a moment to celebrate this win. Every page you've turned here is one step closer to owning the actual test. Now, carry this momentum with you as we move into the Math section, where the same blueprint—stretch, warm-up, drills, and scrimmage—will get you game-ready.

Bottom line: You didn't just finish the English section. You mastered the playbook. On test day, you'll know how to stay calm, stay confident, and stay in control.

Next Steps: How to Keep Preparing

You've completed the scrimmages—great work. But your prep doesn't stop here. There are different ways to continue sharpening your ACT skills. Below, you'll find the Focus Areas for English. It highlights which topics tend to show up most on the test. Pair that with one of three strategies to guide your study plan:

1. **Double Down on Your Strengths**

   Keep practicing what you're already strong at. This locks in easy points and helps you move quickly on test day, freeing time for harder questions.

2. **Prioritize High-Frequency Topics**

   Focus on what shows up most often. Even if you're not perfect, being solid here guarantees a higher return for your effort.

3. **Tackle Your Weaknesses**

   If there's an area that consistently slows you down, spend time there. Improving even a little can boost your confidence and raise your score.

No single approach is "best"—the right choice depends on what you think is best (you know yourself better than anyone else) and how much time you have before test day.

English Focus Areas

The ACT English test covers two main areas: Usage/Mechanics (grammar, punctuation, sentence structure) and Rhetorical Skills (style, organization, and strategy). Key test areas within these categories include Run-ons and Fragments, Subject-Verb Agreement, Pronoun Agreement, proper Punctuation (especially commas), and identifying Wordiness or improving Sentence Style.

Usage and Mechanics

This section focuses on the technical correctness of language.

Sentence Structure: Look for run-on sentences, sentence fragments, faulty parallel structure, and problems with coordinating and subordinating clauses.

Punctuation: Master the correct use of commas, semicolons, colons, dashes, and apostrophes.

Grammar and Usage: Key topics include subject-verb agreement, verb tenses, pronoun agreement and case, adjective and adverb errors, and idiomatic expressions.

Rhetorical Skills:

This section tests your ability to effectively and coherently express ideas in writing.

Strategy: Questions ask how to best achieve an author's goal, whether an idea should be added or deleted, and how to make an argument clearer.

Organization: This involves understanding the logical flow of ideas within a passage, including transitions between paragraphs and sentences.

Style: Evaluate the clarity, conciseness, and tone of writing, identifying and removing wordiness or redundancy and choosing the best word choice.

Key Skills to Focus On

- **Conventions of Standard English:**

This is the largest category and focuses on the rules of grammar, syntax, and mechanics.

- **Production of Writing:**

Focuses on how a passage is developed and the strategies used to convey the author's message.

- **Knowledge of Language:**

Evaluates an understanding of rhetoric, including style, organization, and strategy.

- **Wordiness and Conciseness:**

Learn to identify and remove unnecessary words to make sentences more direct.

From here, you can decide what to do: strengthen your strong spots, focus on what the test throws at you most often, or chip away at your weak areas. **The best prep plan is the one that fits your timeline before test day.**

1. Correct Answer: B / Explanation: 'It's' means 'it is,' but here the possessive 'its' is correct. The tail belongs to the dog.
2. Correct Answer: A / Explanation: All items must be in the same grammatical form. 'Dancing, swimming, and biking' keeps the list parallel.
3. Correct Answer: A / Explanation: 'Because it was raining' is the most direct and concise way to express the idea.
4. Correct Answer: B / Explanation: 'Neither...nor' is the correct correlative pair. 'Coach' is singular, so 'was' is correct.
5. Correct Answer: C / Explanation: This is a comma splice. Use a semicolon, period, or a conjunction with a comma to connect the independent (complete thought) clauses.
6. Correct Answer: C / Explanation: This is a comma splice. Use a semicolon, period, or a conjunction with a comma to connect the independent (complete thought) clauses.
7. Correct Answer: D / Explanation: 'Each of the players' is singular because of "Each", so 'has' is the correct verb form.
8. Correct Answer: B / Explanation: 'Nevertheless' signals contrast, which fits this sentence: classical vs rock.
9. Correct Answer: D / Explanation: This is a dangling modifier. The sentence must make it clear that the dog was barking.
10. Correct Answer: C / Explanation: Avoid redundancy. 'The injured athlete could not play' is clearest.
11. Correct Answer: B / Explanation: 'Literally' is often misused. 'Practically' is more appropriate here.
12. Correct Answer: C / Explanation: The future perfect tense 'will have graduated' is correct for an action completed before a future time.
13. Correct Answer: A / Explanation: This is a list. Commas should separate all items clearly, including before 'and.'
14. Correct Answer: B / Explanation: 'Currently' and 'at this point in time' are redundant. 'We are considering the options' is most concise.
15. Correct Answer: D / Explanation: Option D streamlines the sentence and avoids awkward structure while keeping meaning intact.
16. Correct Answer: A / Explanation: The past tense is needed to match 'yesterday' — 'walked' and 'met' are correct.
17. Correct Answer: A / Explanation: Redundant phrases are removed for clarity and conciseness.
18. Correct Answer: B / Explanation: 'Team' is singular and takes 'is' as the correct verb.
19. Correct Answer: A / Explanation: Nonessential clause 'which are old' should be set off with commas.
20. Correct Answer: A / Explanation: Avoid redundancy — 'The reason is because' is repetitive.
21. Correct Answer: B / Explanation: The modifier 'Running through the woods' must refer to 'he,' not 'the trees.' B places the modifier correctly.

22. Correct Answer: A / Explanation: 'Nevertheless' shows contrast, making it the correct transition for continuing despite fatigue.
23. Correct Answer: B / Explanation: Option B is the clearest and most concise, removing unnecessary words without losing meaning.
24. Correct Answer: C / Explanation: Correct use of the Oxford comma: items in a list should be separated by commas, including before 'and.'
25. Correct Answer: B / Explanation: 'Committee' is a collective noun and takes a singular pronoun — 'its' is correct.
26. Correct Answer: C / Explanation: Subject pronouns 'He' and 'I' are correct for the subject position.
27. Correct Answer: D / Explanation: All items are nouns — 'attention,' 'punctuality,' and 'organization' — keeping the structure parallel.
28. Correct Answer: C / Explanation: 'Unsuccessful' is formal and academic, unlike slang such as 'kinda' or 'total fail.'
29. Correct Answer: A / Explanation: Use 'were' in hypothetical or contrary-to-fact statements: 'If I were taller...'
30. Correct Answer: B / Explanation: The bone belongs to one dog — singular possessive form is 'dog's bone.'
31. Correct Answer: A / Explanation: Option A is direct and eliminates unnecessary words for clarity.
32. Correct Answer: B / Explanation: Past perfect ('had started') is used to show something occurred before another past event.
33. Correct Answer: C / Explanation: Option C correctly joins two independent clauses with a comma and a coordinating conjunction.
34. Correct Answer: A / Explanation: Formal grammar requires 'his or her' with a singular indefinite pronoun like 'everyone.'
35. Correct Answer: D / Explanation: When using 'neither/nor,' the verb agrees with the subject closest to it — in this case, 'coach' (singular), so 'is' is correct.
36. Correct Answer: C / Explanation: Option C maintains correct parallel structure with 'not only...but also' without repeating 'was.'
37. Correct Answer: B / Explanation: A comma is needed after an introductory phrase: 'After dinner, we had dessert.'
38. Correct Answer: B / Explanation: B clearly shows that the pancakes are covered in syrup. Other choices misplace or dangle the modifier.
39. Correct Answer: C / Explanation: C eliminates redundancy. 'The reason why... is because...' is wordy and repetitive.

40. Correct Answer: B / Explanation: B is the most concise and clear version, avoiding awkward or redundant phrasing.
41. Correct Answer: B / Explanation: Colons must follow a complete sentence when introducing a list. B does so properly.
42. Correct Answer: C / Explanation: Subject pronouns ('She' and 'I') are used correctly as the subject of the sentence.
43. Correct Answer: D / Explanation: 'Who is a doctor' is nonessential information and should be set off with commas.
44. Correct Answer: A / Explanation: 'Complete' is the correct word meaning to finish something.
45. Correct Answer: B / Explanation: 'Team' is a singular noun, so it requires 'has' and 'its' for subject-verb and pronoun agreement.
46. Correct Answer: B / Explanation: 'End result' and 'final outcome result' are redundant. Option B is the most concise and clear.
47. Correct Answer: C / Explanation: All items use gerunds ('-ing' verbs), maintaining parallel structure.
48. Correct Answer: B / Explanation: 'List' is the subject, and it is singular. So, the verb must be 'is.'
49. Correct Answer: B / Explanation: With 'neither/nor,' the verb agrees with the closest subject — 'dad,' which is singular. So, 'likes' is correct.
50. Correct Answer: C / Explanation: Option C clearly reinforces the argument and ties back to the main idea with a persuasive tone.
P1.1: Correct Answer: B / Explanation: 'Concerns' is plural, so the correct verb is 'are.'
P1.2: Correct Answer: C / Explanation: The correct phrase is 'consist of.' 'Comprise of' is incorrect usage.
P1.3: Correct Answer: B / Explanation: Subject-verb agreement and parallel structure require 'increases' and 'builds.'
P1.4: Correct Answer: C / Explanation: Passive voice requires 'be submitted' to match 'will likely be.'
P2.1: Correct Answer: C / Explanation: 'That attracts' is the correct relative clause for the subject 'event.'
P2.2: Correct Answer: A / Explanation: The sentence is fine as is with past tense 'said' to match the narrative.
P2.3: Correct Answer: B / Explanation: Singular subject 'it' requires a singular verb 'brings.'
P2.4: Correct Answer: B / Explanation: Collective noun 'group' takes singular verb 'was preparing.'
P3.1: Correct Answer: D / Explanation: 'She constructed herself' is concise and grammatically correct.
P3.2: Correct Answer: B / Explanation: Correct possessive form is 'her ability'; 'she's' is incorrect here.
P3.3: Correct Answer: B / Explanation: Singular subject 'project' requires 'demonstrates' for subject-verb agreement.
P3.4: Correct Answer: C / Explanation: 'Had been awarded' fits best for an action completed by a specific past time.
P4.1: Correct Answer: A / Explanation: 'Turning her hobby into a business' is correct idiomatic usage.

P4.2: Correct Answer: B / Explanation: The correct phrase is 'should have started'—'should of' is incorrect.

P4.3: Correct Answer: B / Explanation: 'One' is the subject and requires the singular verb 'was'.

P4.4: Correct Answer: A / Explanation: 'Became popular' is the best and clearest expression of the completed action.

P5.1: Correct Answer: A / Explanation: 'Dreamed of traveling' is the correct and idiomatic phrasing for past tense.

P5.2: Correct Answer: B / Explanation: All items in the list must follow the same verb form: 'eating… attending.'

P5.3: Correct Answer: C / Explanation: 'Lost' is the correct past tense form; 'loosed' and 'loose' are incorrect in this context.

P5.4: Correct Answer: B / Explanation: The correct infinitive phrase is 'how to appreciate'.

1. A. Unusual is neutral, appropriate for a nonfiction informative passage. The other choices (weird, strange, oddball) are too casual or informal.

2. H. Was matches with the singular word "frame" and is the correct tense (past tense).

3. A. It provides specific details that demonstrate how the project was carried out, including information about design and funding, thereby making the project's development concrete. Without it, the passage lacks specifics.

4. J. A semicolon correctly connects two independent clauses with a transitional adverb (*however*) starting the second independent clause and correctly has the comma after "however".

5. B. Signals contrast and resolution after describing setbacks. The others do not fit the logic.

6. J. The phrase "which drew the attention of local news outlets" is nonrestrictive. Nonrestrictive clauses add extra information that isn't essential to the meaning of the sentence. This nonrestrictive clause is introduced by which and is set off by a comma.

7. D. "Drew" is the clearest synonym for "attracted attention" in this context. The others (pulled, lured, gathered) carry unintended or awkward connotations.

8. H. parentheses, commas, and dashes are all acceptable for setting off parenthetical elements (non-essential information, so you can also delete them). Quotation marks are not.

9. B. Clarifies the intended meaning of *"beyond one bench."* The other answers are exaggerated (A), illogical (C, D), or inaccurate.

10. J. At the conclusion, this sentence reinforces the project's broader significance. Placing it earlier would interrupt the narrative of events.

11. D. A colon (:) is used to introduce a list, an explanation, or extra information that directly relates to what comes before it. Quick Trick for ACT: If what comes after the colon completes or explains the thought before it, the colon is correct. What comes after must not be random—it must connect directly. What comes

before a colon must be a complete sentence. What comes after the colon doesn't have to be an independent clause.

12. G. "Murals are messy, and the paint will fade" is the clearest, most concise option. A and C are wordy; D is grammatically awkward.

13. A. It shows the students' nervousness, adding perspective before the meeting. G is incorrect — it doesn't explain the sketch content. H and J incorrectly claim repetition or off-topic focus.

14. H. In the sentence: "The students raised money by hosting a pancake breakfast and applied for a grant from the town's cultural arts fund." The subject "The students" is shared across both actions: (1) raised money by hosting a pancake breakfast, (2) applied for a grant. Since the second part **does not have its own subject** stated again, this is a **compound predicate** (two verbs/actions tied to the same subject). No comma is needed. Quick Trick for ACT: If one subject is doing two actions, **no comma** before *and.* If each action has its own subject, you do need a comma and FANBOY.

15. D. Best shows contrast between expectations (fun painting) and reality (hard physical work).

16. H. Most precise and formal: "mixing paint so that the colors matched their designs." F, G, and J are vague or awkward.

17. A. Specifically shows how the community supported the project. B, C, and D are unrelated to community involvement.

18. F. No pause or interruption is needed after "picture."

19. A. It demonstrates the impact of the mural on community members. Adds emotional impact at the unveiling. B is wrong — it doesn't repeat details. C and D incorrectly dismiss the value of community reactions.

20. H. placing the sentence at [C] creates a smooth transition from gaining approval to raising funds ("The students raised money…"). At [A], the idea is too early, before doubts arise; at [B], it would interrupt the logic of scheduling a meeting. Option D is incorrect because this transitional sentence improves coherence.

ACT English tests skills in grammar, punctuation, sentence structure, and rhetorical skills, which are crucial for jobs that require clear and effective communication. A strong score indicates an ability to write and edit with precision, making it valuable for careers in writing, law, communications, and education.

10 jobs that require ACT English skills

1. **Lawyer:** A strong command of language is essential for success in law. The work involves deep analysis of complex legal texts, drafting legal documents with impeccable grammar, and constructing clear and persuasive arguments both in writing and speech.

2. **Editor:** Editors oversee written material for publication across various media, including books, magazines, websites, and marketing collateral. They need exceptional attention to detail to catch errors in spelling, punctuation, and grammar, while also ensuring the text has a consistent tone and style.

3. **Journalist:** This profession demands strong skills in writing, organization, and rhetoric to accurately and effectively report news and tell stories. Journalists must clearly communicate complex information while adapting their style to suit different audiences and formats.

4. **Technical Writer:** Technical writers create clear, concise instructional materials, such as user manuals, how-to guides, and assembly instructions. Their job is to take complex subjects and make them easy for a layperson to understand, which requires a mastery of language.

5. **Copywriter:** As a marketing role, copywriting involves crafting persuasive text for advertising and promotional materials. Copywriters must be skilled in using language strategically to influence an audience and promote products or services effectively.

6. **Teacher/Professor:** English skills are the foundation of a career in education, particularly for those who teach English, literature, or writing at the high school

or university level. Educators must possess and impart a strong command of grammar, usage, and rhetorical skills.

7. **Public Relations Manager:** This role involves managing an organization's public image through various forms of communication. Public relations managers write press releases, create social media content, and draft speeches, all of which require impeccable writing skills and an understanding of audience and tone.

8. **Grant Writer:** Grant writers are essential for non-profit organizations seeking funding. They must write compelling and detailed grant proposals that outline the organization's mission and needs. This requires precise, professional, and persuasive writing abilities.

9. **Librarian:** A career as a librarian requires strong research and organizational skills. While often needing a master's degree, an undergraduate English background is excellent preparation for curating information and assisting people with their research and reading needs.

10. **Content Strategist/Manager:** In the digital age, this role is critical for creating and implementing a consistent and effective content strategy for a website or company. These professionals need excellent writing and editing skills to plan, create, and manage content across multiple platforms.

MATH

Chapter 4: Math Stretch

1. What is the value of x in the equation $2x + 3 = 11$?

A. 3

B. 4

C. 5

D. 6

2. If a triangle has angles measuring 35° and 65°, what is the measure of the third angle?

A. 80°

B. 85°

C. 90°

D. 100°

3. What is the slope of the line that passes through the points (2, 3) and (6, 11)?

A. 2

B. 3

C. 4

D. 2.5

4. Simplify: $3 (x - 2) + 4x$.

A. $7x - 6$

B. $x + 6$

C. $3x - 2$

D. $7x + 6$

5. If $f(x) = x^2 - 3x + 2$, what is $f(2)$?

A. 0

B. 2

C. 4

D. 6

6. What is the value of $3^2 + 4^2$?

A. 12

B. 25

C. 7

D. 5

7. Simplify: $5x – 2x + 7$

A. $3x + 7$

B. $7x – 2$

C. $5x + 5$

D. $3x – 7$

8. What is the slope of the line $y = −2x + 3$?

A. 2

B. −2

C. 3

D. −3

9. Solve for x: $2(x − 1) = 10$

A. 4

B. 5

C. 6

D. 7

10. A triangle has angles measuring 35° and 75°. What is the measure of the third angle?

A. 70°

B. 80°

C. 85°

D. 65°

11. Factor: $x^2 - 5x + 6$

A. $(x - 2)(x - 3)$

B. $(x + 2)(x - 3)$

C. $(x - 1)(x - 6)$

D. $(x - 2)(x + 3)$

12. If $f(x) = 2x^2$, what is $f(-3)$?

A. −18

B. 18

C. 9

D. −9

13. What is the midpoint between (2, 3) and (6, 7)?

A. (4, 5)

B. (8, 10)

C. (3, 4)

D. (2, 2)

14. Solve: $x^2 - 4x = 0$

A. $x = 2$

B. $x = 0$ or $x = 4$

C. $x = 0$ or $x = -4$

D. $x = 1$ or $x = 4$

15. What is 40% of 60?

A. 24

B. 30

C. 36

D. 40

16. If a rectangle has a length of 10 and a width of 4, what is the perimeter?

A. 14

B. 40

C. 28

D. 25

17. What is the value of $\frac{3}{4} \times \frac{8}{9}$?

A. 2/3

B. 2/5

C. 3/5

D. 6/7

18. If $x + 2 = 3x - 4$, what is x?

A. 3

B. 2

C. 4

D. −3

19. Find the average of 4, 7, 9, and 10.

A. 7.5

B. 8

C. 7

D. 6.5

20. If y = 2x and x = 5, what is y?

A. 10

B. 7

C. 8

D. 12

21. A shirt costs $20 and is on sale for 25% off. What is the sale price?

A. $5

B. $15

C. $20

D. $25

22. Find the area of a circle with radius 3.

A. 9π

B. 6π

C. 12π

D. 18π

23. What is the median of the data set: 4, 8, 6, 2, 10?

A. 6

B. 5

C. 7

D. 8

24. What is the area of a triangle with a base of 10 and a height of 5?

A. 25

B. 30

C. 50

D. 20

25. What is the height of a triangle with a base of 10 and an area of 50?

A. 5

B. 10

C. 15

D. 20

26. What is the next number in the pattern: 2, 4, 8, 16, __?

A. 20

B. 24

C. 30

D. 32

27. What is the value of $X^4 \div X^2$?

A. x^1

B. x^2

C. x^6

D. x^8

28. The ratio of boys to girls is 2:3. If there are 20 students, how many are girls?

A. 8

B. 10

C. 12

D. 15

29. Find the slope between (2,3) and (4,7)

A. 2

B. 3

C. 4

D. 1

30. The ratio of boys to girls is 4:6. If there are 20 students, how many are boys?

A. 8

B. 10

C. 12

D. 15

31. Find the slope between (2,3) and (4,7)

A. 2

B. 3

C. 4

D. 1

32. If $\log_{10}(100) = x$, what is x?

A. 1

B. 2

C. 10

D. 100

33. Solve for x: $2x - 5 = 9$

A. 2

B. 3

C. 7

D. 8

34. Which of these is a factor of $x^2 - 3x + 2$?

A. $(x - 1)$

B. $(x + 1)$

C. $(x + 2)$

D. $(x + 3)$

35. What is the value of $3^2 \times 3^3$?

A. 243

B. 27

C. 81

D. 243

Chapter 5: Math Warm-up and Drills

As we head into the math warm-up phase, it's time to open our playbook. Think of this as your **pre-game strategy session**—the place where you scout out the types of problems, the pace of the game, and the traps the test-makers love to set.

Just like in basketball or football, you wouldn't step on the court or field without knowing your opponent's style. The ACT Math section has its own tendencies. They'll hit you with straightforward questions to get you comfortable, then sneak in curveballs that test your timing, accuracy, and composure.

Here's the good news: once you know the plays, the "opponent" isn't so tough. The math section follows patterns—algebra, geometry, functions, statistics—and once you've seen how they like to mix them up, you can predict what's coming. Preparation turns those traps into easy points on the scoreboard.

Think of every drill in this section as a **practice rep**. You're running plays, sharpening your moves, and building the confidence to attack the real game. By the time test day arrives, you'll already know the playbook—and you'll be ready to execute it.

Now let's hit the floor, run through the drills, and lock in the math game plan.

SCOUTING REPORT
KNOW THE PLAY.
MASTER THE GAME.
MOORE SUBSTANCE
TEST PREP

Half of the ACT Math section is built on pre-algebra and basic algebra. That's good news—because if you sharpen those skills, you already give yourself a strong chance of scoring at least a 19.

Here's the catch: most students learn pre-algebra back in middle school. By the time junior year rolls around, it's been years since those concepts got real practice. Add in differences in teachers, classrooms, and whether the material ever made it into long-term memory, and it's no wonder this part can trip people up.

Math on the ACT also has a unique twist—it's the only section that gets harder as you go. That means the earlier problems (pre-algebra, basic/intermediate algebra) are your best chance to pick up points quickly. If geometry is your strong suit, lean into that too—it's all about stacking points where you're strongest so you can save time for the tougher problems.

And don't forget, formulas are your secret weapon. Memorizing the right ones can turn a 2-minute question into a 15-second win.

This section gives you the formulas, strategies, and pro test tips you'll need to boost your score and step onto solid ground for the rest of the Math test.

Algebra & Arithmetic

Slope = $(y_2 - y_1) / (x_2 - x_1)$

Slope-intercept form: $y = mx + b$

Quadratic formula: $x = [-b \pm \sqrt{(b^2 - 4ac)}] / 2a$

FOIL: $(a + b)(c + d) = ac + ad + bc + bd$ [you can also use the box method]

Distance formula: $\sqrt{[(x_2 - x_1)^2 + (y_2 - y_1)^2]}$

Midpoint formula: $((x_1 + x_2)/2 , (y_1 + y_2)/2)$

Exponent Rules: x□ * x□ = x□$^{+}$□ ; (x□)□ = x□□ ; $x^0 = 1$

Geometry

Area of triangle = ½ × base × height

Area of circle = πr^2

Circumference of circle = $2\pi r$

Pythagorean Theorem: $a^2 + b^2 = c^2$

Area of a rectangle = length × width

Area of trapezoid = ½ × ($base_1$ + $base_2$) × height

Area of a Parallelogram = A = base x height

Volume of rectangular prism = l × w × h

Volume of cylinder = $\pi r^2 h$

Equation of a circle = $(x - h)^2 + (y-k)^2 = r^2$

Trigonometry

SOHCAHTOA (sin = opp/hyp, cos = adj/hyp, tan = opp/adj)

$\sin^2\theta + \cos^2\theta = 1$

Basic unit circle values (0°, 30°, 45°, 60°, 90°) **0 radians (0°):** (1, 0), **π/6 radians (30°):** (√3/2, 1/2), **π/4 radians (45°):** (√2/2, √2/2), **π/3 radians (60°):** (1/2, √3/2), **π/2 radians (90°):** (0, 1).

Probability & Statistics

Mean = sum of values / number of values

Median = middle number (or average of two middle numbers)

Probability = desired outcomes / total outcomes

Top Strategies for ACT Math

- Use your calculator wisely – don't reach for it out of habit when you can use your brain. Memorize your multiplication table along with perfect squares, perfect cubes, and these Pythagorean Triples: 3-4-5, 5-12-13, 7-24-25, because they are commonly used and knowing them will save you time from not having to use $a^2 + b^2 = c^2$. When you do use your calculator, be careful: incorrect input yields undesired output.
- Work backwards on multiple-choice questions by plugging in answer choices. Start in the middle when plugging in choices, so you know whether to proceed with higher or lower numbers. All answer choices start with the lowest number and proceed in ascending order.
- Eliminate extreme answers and outliers to save time.
- Draw diagrams to visualize problems, especially for geometry problems, when an image isn't given. When an illustration is given, always use it. There's a myth that photos are not fit to scale, but they are. The instructions say, "Illustrative figures are **NOT** necessarily drawn to scale." Key word "necessarily. They try to trick you with the "NOT" and hope you overlook "necessarily" because they will help you solve them without using any math in some instances.
- Don't spend too much time on difficult ones during your first pass through. The goal is to answer every question (They're all worth the same number of points). Math can get progressively more challenging, BUT they sprinkle in easier ones near the end as well. They're just hoping you're burnt out or have given up by then.

- The more reading you must do on a word problem, the easier it is. They know you had to spend time reading it, so they make the math a little easier. Be sure to find exactly what they're asking you to solve at the end of the question. Don't get caught up in the lengthy wording at the start or trying to answer what's familiar (what's normally asked for that type of question).
- Use estimation when exact values aren't necessary. For multiple variables, plug in your own numbers. When plugging in your own numbers to solve, avoid 0 and 1 (multiplication reasons) and keep your numbers low so that math is simpler (2, 3, 4, and 5 at most). Once solved, plug those same numbers in for the variables in the answer choices. Whichever yields the same result as your initial answer is the correct one.
- Don't forget your basics: order of operations (PEMDAS), fractions, and proportions.
- Know common math vocabulary like 'product,' 'sum,' 'difference,' 'addition,' 'integer,' 'range,' 'mean,' 'median,' and 'mode.'
- Don't forget your basics: order of operations (PEMDAS), and how to solve fractions, percents, decimals, and ratios and proportions. Always double-check units and conversions.

As previously mentioned, Strategies are only as good as the practice behind them. To sharpen your skills, let's move into some Math word problem "drills."

1. Arlette's favorite punch recipe requires 4 cups of juice and makes 10 servings. She decides to modify the recipe by using 7 cups of juice and increasing all other ingredients proportionally. How many total servings will the modified recipe make?

A. 14.5

B. 16

C. 17.5

D. 19

E. 21.5

2. The 40-member Science Club is meeting to select a representative for the district competition. The members decide that the representative, who will be chosen at random, CANNOT be any of the 5 officers of the club. What is the probability that Jordan, who is a member of the club but NOT an officer, will be chosen?

A. 0

B. $\frac{1}{35}$

C. $\frac{1}{40}$

D. $\frac{5}{40}$

E. $\frac{1}{5}$

3. The City Gym charges a $50 monthly membership fee plus $4 for each yoga class taken in that month. The Community Gym charges a $40 monthly fee plus $6 for each yoga class taken in that month. For how many yoga classes in one month will the total charges from both gyms be equal?

F. 2

G. 3

H. 5

J. 6

K. 10

4. The average of 5 distinct quiz scores is equal to the median of the 5 scores. The sum of the 5 scores is 475. What is the sum of the 4 scores that are **not** the median?

F. 325

G. 360

H. 375

J. 380

K. 420

5. In Bankhead County, the fine for running a red light is $25 for each second the driver enters the intersection after the light turns red. In Bankhead County, Aaron was fined $300 for entering the intersection late. At how many seconds after the light turned red did Aaron enter the intersection?

F. 10

G. 11

H. 12

J. 13

K. 14

6. Orlando is deciding whether to buy a season pass to his university's 20 home basketball games this year. A single-game ticket costs $16, and a season pass costs $180 and admits him to any home game at no additional cost. What is the minimum number of home games Orlando must attend this season for the season pass to cost less than buying single-game tickets?

A. 8

B. 10

C. 11

D. 12

E. 20

7. The lengths of the 2 legs of right triangle ΔXYZ shown below are given in inches. The midpoint of the hypotenuse $\overline{XY}$ is how many inches from X?

A. 10

B. 13

C. 15

D. 17

E. 20

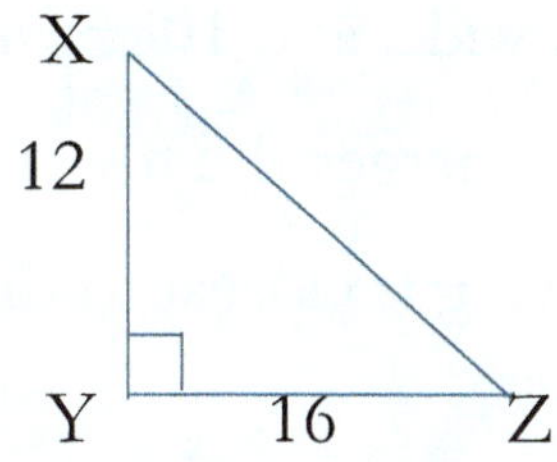

Triangle midpoint diagram

8. The sum of 2 positive numbers is 119. The lesser number is 9 more than the square root of the greater number. What is the value of the greater number minus the lesser number?

F. 19

G. 66

H. 81

J. 91

K. 121

9. At a local fundraiser, Jossie will play a game in which she flips a quarter, a half-dollar, and a dollar coin at the same time. She will be awarded 4 points for each coin that lands heads up. Let the random variable y represent the total number of points Sarah earns on any flip of the coins. What is the expected value of y?

A. 2

B. 3

C. 6

D. 9

E. 12

10. Aryiana wants to find the volume of a solid toy car. She fills a rectangular container 12 cm long, 5 cm wide, and 10 cm high with water to a depth of 3 cm. Aryiana then completely submerges the toy car in the water. The water level rises to 5.4 cm. Which of the following is closest to the volume, in cubic centimeters, of the toy car?

A. 120

B. 150

C. 180

D. 210

E. 240

11. A rectangular garden has dimensions 18 feet by 27 feet. What is the area of the garden in square *yards*?

A. 36

B. 48

C. 54

D. 81

E. 162

12. A bakery made a large cake. Emma ate only $\frac{3}{8}$ of it and shared the rest equally among her 5 cousins. What fraction of the whole cake did each cousin receive?

F. $\frac{3}{40}$

G. $\frac{1}{8}$

H. $\frac{5}{32}$

J. $\frac{3}{5}$

K. $\frac{5}{8}$

13. The equation $|3x-9|+4=10$ has 2 solutions. Those solutions are equal to the solutions to which of the following pairs of equations?

F. $3x-9=6$

$-3x-9=6$

G. $3x-9=6$

$-(3x-9)=6$

H. $3x-9=10$

$-(3x-9)=10$

J. $3x - 9 = 14$

$-(3x - 9) = 14$

K. $3x - 9 = 2$

$-(3x - 9) = 2$

14. In the standard (x,y)(x,y)(x,y) coordinate plane, what is the slope of the line given by the equation

$$5x = 2y - 10?$$

A. $-\frac{2}{5}$

B. $\frac{2}{5}$

C. $\frac{5}{2}$

D. 2

E. 5

15. What is the sum of the solutions of the 2 equations below?

$$6x = 18$$

$$3y - 5 = 16$$

A. $\frac{13}{2}$

B. 10

C. 12

D. $\frac{35}{6}$

E. 15

16. For which of the following conditions will the sum of integers *a* and *b* **always** be an odd integer?

F. *a* is an odd integer.

G. *b* is an odd integer.

H. *a* and *b* are both odd integers.

J. *a* and *b* are both even integers.

K. *a* is an odd integer, and *b* is an even integer.

17. The average of a list of 4 numbers is 92.0. A new list of 4 numbers has the same first 3 numbers as the original list, but the fourth number in the original list is 76, and the fourth number in the new list is 100. What is the average of this new list of numbers?

A. 94

B. 95

C. 96

D. 97

E. 98

18. For what positive real value of k, if any, is the determinant of the matrix

$$\begin{bmatrix} k & 7 \\ 8 & k \end{bmatrix}$$

equal to k?

(Note: The determinant of matrix $\begin{bmatrix} a & b \\ c & d \end{bmatrix}$ equals ad – bc.)

A. 4

B. 6

C. 7

D. 8

E. There is no such value of k

19. The list of numbers 82, 70, 60, X, Y, 30 has a median of 50. The mode of the list of numbers is 30. To the nearest whole number, what is the mean of the list?

A. 44

B. 48

C. 50

D. 52

E. 56

20. What is the slope of the line through (−3, 4) and (3,−5) in the standard (x , y) coordinate plane?

A. −1

B. 1

C. $-\frac{3}{2}$

D. $\frac{3}{2}$

E. −3

21. In the standard (x ,y) coordinate plane, what is the midpoint of the line segment that has endpoints (6, 10) and (2, −6)?

A. (3, 2)

B. (4, 2)

C. (2, 3)

D. (5, 0)

E. (8, −8)

22. What is the sum of the first 4 terms of the arithmetic sequence in which the 6th term is 8 and the 10th term is 13?

A. 10.5

B. 14.5

C. 18

D. 21.25

E. 39.5

23. The 5th and 6th terms of an arithmetic sequence are 20 and 26, respectively. What is the 50th term of the sequence?

A. 274

B. 284

C. 290

D. 296

E. 304

24. If x and y are real numbers such that $x > 1$ and $y < -1$, then which of the following inequalities must be true?

A. $\frac{x}{y} > 1$

B. $|x|^2 > |y|$

C. $\frac{x}{3} - 5 > \frac{y}{3} - 5$

D. $x^2 + 1 > y^2 + 1$

E. $x^{-2} > y^{-2}$

25. A portion of a city has the shape and dimensions of the blocks given below. All the borders run either north-south or east-west. Three friends decide to meet. One is at point D, another at point B, and the third at point A If they want to meet halfway between point D and B, which of the following is the location of the meet-up from point A?

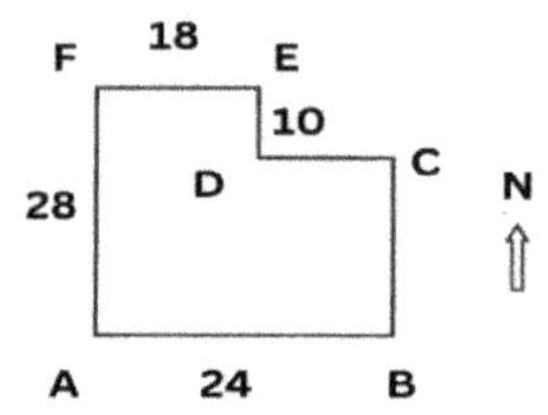

A. 19 blocks east and 9 blocks north

B. 18 blocks east and 5 blocks north

C. 21 blocks east and 9 blocks north

D. 21 blocks east and 10 blocks north

E. 24 blocks east and 18 blocks north

Chapter 5: Math Scrimmage

You've stretched. You've warmed up. You've run the drills.

Now it's time to **scrimmage** — where practice meets pressure.

This isn't just any practice set — these are math questions designed in the style of the ACT— **authentic** in timing, format, and difficulty.) **ACT math questions** under **real timing** with **real challenge**. Here's your chance to test your pacing, sharpen your problem-solving, and learn how to manage the clock when it matters most.

Remember — scrimmages aren't about perfection. They're about learning to play under **game-time conditions**. Every mistake is just film for us to study later, and every win helps build momentum, so lock in, stay composed, and trust your training.

How to Time Your Scrimmage

- **Option 1: Full 25 Questions (Recommended)**

  □ **28 minutes total** — the same proportional time you'll have on the new **Enhanced ACT**.

  This will give you a true feel for test-day pacing and stamina.

- **Option 2: Break It into Sets**

  □ **10 questions within 11 minutes.**

- **15 questions within 17 minutes.**

  Pause your timer after each set to review your answers and strategies.

- **Trust your training** — stick to the methods you've practiced.
- **Don't get stuck** — flag tough questions and come back if time allows.
- **Keep the pace steady** — that's your edge with the new ACT timing.
- This scrimmage is about building speed, accuracy, and confidence — the three keys to your best ACT math score.

1. $|5 - 2| - |2 - 5| = ?$

A. –3

B. –6

C. 0

D. 3

E. 6

2. If 30% of a given number is 12, then what is 5% of the given number?

F. 2

G. 6

H. 12

J. 15

K. 20

3. The figure below is composed of square BCDE and an equilateral triangle Δ ABE. The length of $\overline{CD}$ is 5 inches. What is the perimeter of ABCDE, in inches?

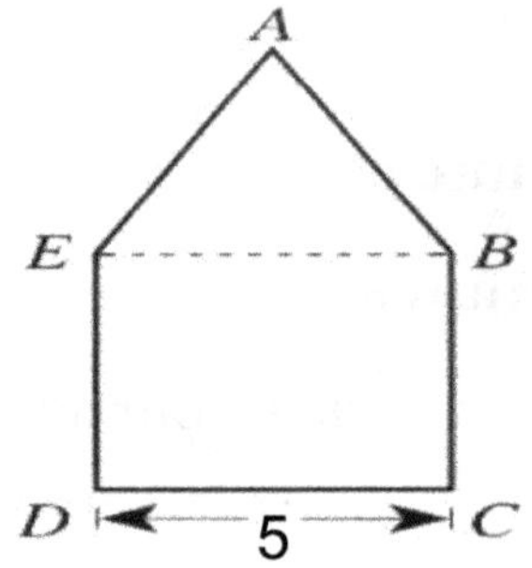

A. 10

B. 15

C. 20

D. 25

E. 30

4. For what value of x is the equation $4^{2x + 9} = 4^{15}$ true?

F. 2

G. 3

H. 4

J. 5

K. 6

5. Let the function f be defined as $f(x) = 5x^2 - 3(2x + 3)$. What is the value of $f(2)$?

A. –2

B. –1

C. 1

D. 2

E. 3

6. The first term is 2 in the geometric sequence 2, −6, 18, −54, …. What is the SEVENTH term of the geometric sequence?

F. –1458

G. –162

H. 162

J. 486

K. 1458

7. The width of a rectangle with an area of 45 square inches is 5 inches. What is the perimeter of the rectangle, in inches?

A. 9

B. 18

C. 28

D. 45

E. 90

8. In the figure below, C is the intersection of $\overline{AD}$ and $\overline{BE}$. If it can be determined, what is the measure of ∠ BAC ?

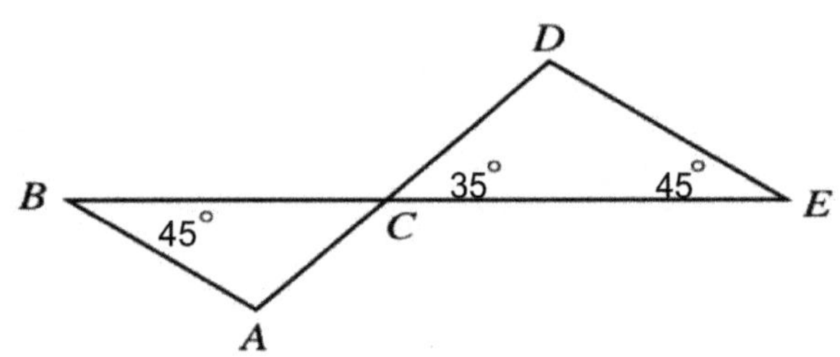

F. 35°

G. 45°

H. 80°

J. 100°

K. Cannot be determined from the given information.

9. The 6 consecutive integers below add up to 333.

X – 1

X – 2

X

X + 1

X + 2

X + 3

What is the value of X ?

A. 51

B. 53

C. 55

D. 57

E. 59

10. Which of the following expressions is equivalent to $\frac{1}{2}y^2$ (12x +4y + 24x − 4y) ?

F. $18xy^2$

G. 36xy

H. $9xy^2 + 24x$

J. $18xy^2 - 2y^3$

K. $6xy^2 + 24x - 2y^3 - 4y$

11. If x, y, and z are the positive integers such that $x^y = a$ and $z^y = b$, then ab = ?

A. xz^y

B. xz^{2y}

C. $(xz)^y$

D. $(xz)^{2y}$

E. $(xz)^{y2}$

12. In the figure below $\Delta\overline{ABC}$ and $\Delta\overline{RST}$ are similar triangles. What is the perimeter of $\Delta\overline{ABC}$?

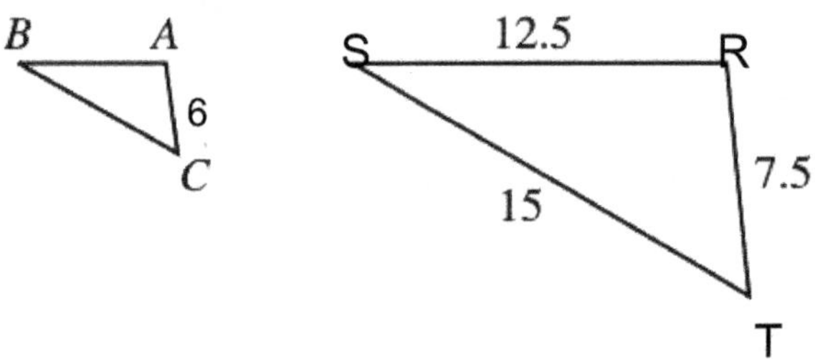

F. 24

G. 28

H. 30

J. 35

K. 40

13. Christa cut a 50-foot-long rope into 2 pieces. The ratio of the lengths of the 2 pieces is 2:3. What is the length, to the nearest foot, of the shorter piece?

A. 10

B. 15

C. 20

D. 25

E. 30

14. For all d > 1, the expression $\frac{4d^4}{4d^6}$ equals:

F. $\frac{1}{2}$

G. $-d^2$

H. d^2

J. $-\frac{1}{d^2}$

K. $\frac{1}{d^2}$

15. A right circular cylinder is shown in the figure below, with dimensions given in centimeters. What is the total surface area of this cylinder, in square centimeters? (Note: The total surface area of a cylinder is given by $2\pi r^2 + 2\pi rh$, where r is the radius and h is the height.)

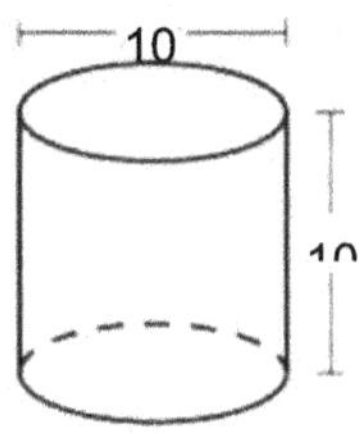

A. 150π

B. 200π

C. 250π

D. 300π

E. 350π

16. For the right triangle $\Delta\overline{EFG}$, what is sinθ ∠F ?

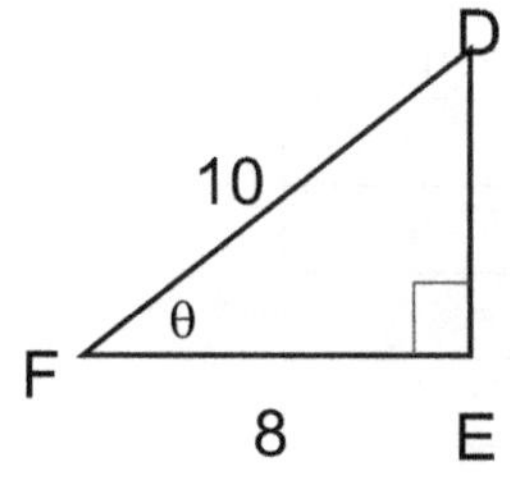

F. $\frac{3}{5}$

G. $\frac{3}{4}$

H. $\frac{4}{5}$

J. $\frac{4}{3}$

K. $\frac{5}{3}$

17. A formula used to compute the current value of a savings account is $A = P(1 + r)^n$, where A is the current value; P is the amount deposited; r is the rate of interest for 1 compounding period, expressed as a decimal; and n is the number of compounding periods. Which of the following is closest to the value of a savings account after 3 years if $15,000 is deposited at 2% annual interest compounded yearly?

A. 15,300

B. 15,560

C. 15,873

D. 15,918

E. 16,300

18.

Use the following information to answer questions 18-20.

200 certified NBA voters voted for the best player over the last decade. It was narrowed down to four players, and all 200 indicated which 1 of the 4 finalists they would vote for. The poll results are presented in the table below.

Player	Number of votes
G. Antetokounmpo	30
Stephen Curry	50
Kevin Durant	40
LeBron James	80

18. What percent of voters voted for Steph Curry?

F. 15%

G. 20%

H. 25%

J. 30%

K. 35%

19. If this voting poll is indicative of how 1,000 of the total certified voters will ultimately vote, which of the following is the best estimate of the votes LeBron would get?

A. 250

B. 300

C. 350

D. 400

E. 450

20. If the information in the table were converted into a circle graph (pie chart), then the central angle of the sector for Durant would measure how many degrees?

F. 64°

G. 72°

H. 80°

J. 84°

K. 88°

21. If $\frac{7\sqrt{11}}{a\sqrt{11}} = \frac{7\sqrt{11}}{11}$ is true, then a = ?

A. -1

B. 1

C. $\sqrt{11}$

D. 11

E. 121

22. Trapezoid ABCD is graphed in the standard (x,y) coordinate plane below. What is the slope of $\overline{CD}$?

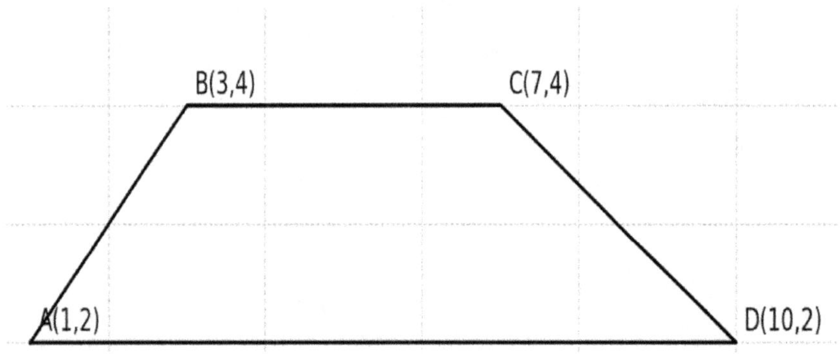

F. $-\frac{2}{3}$

G. $-\frac{3}{2}$

H. $\frac{2}{3}$

J. 1

K. $\frac{3}{2}$

23. A student can read 5 three-hundred-page books in 3 days. At this rate, how many three-hundred-page books can a student read in 3 + d days?

A. $\frac{5}{3} + d$

B. $\frac{5}{3} + \frac{d}{3}$

C. $\frac{5}{3} + \frac{5}{3d}$

D. $5 + \frac{d}{3}$

E. $5 + \frac{5d}{3}$

24. You can find the volume of an irregularly shaped solid object by completely submerging it in water and calculating the volume of water the object displaces. You completely submerge a solid, irregularly shaped object in a rectangular tank with a base of 50 centimeters by 30 centimeters, filled with water to a depth of 25 centimeters. The object sinks to the bottom, and the water level goes up 0.35 centimeters. What is the volume, in cubic centimeters, of the object?

F. 125

G. 250

H. 325

J. 450

K. 525

25. A container is $\frac{1}{8}$ full of water. After 20 cups of water are added, the container is $\frac{3}{4}$ full. What is the volume of the container, in cups?

A. 24

B. 27

C. 30

D. 32

E. 38

Congratulations — you just powered through the entire Math section scrimmage! That's no small feat. You battled through equations, graphs, geometry, word problems, and data sets—all while keeping the clock in check.

Here's What This Means for You

- You've built stamina. Endurance matters on test day, and you just proved you can handle the pace of a full 45-question enhanced section.
- You've tested your strategies. From eliminating wrong answers quickly to plugging in numbers and using shortcuts, you've seen where your approach shines—and where you may need to sharpen it.
- You've learned under pressure. The ACT isn't just about math skills; it's about managing time and stress. Practicing with real timing conditions makes you calmer and more confident when it counts.

Next Steps

Take a moment to celebrate this win. Every problem you solved, skipped, or came back to is valuable data for us to study. Review any patterns in your mistakes—whether it's setting up equations, interpreting graphs, or pacing—and use those insights to strengthen your game plan.

This scrimmage wasn't just about getting answers right; it was about mastering process and performance. On test day, you'll know how to stay calm, stay sharp, and stay in control.

Next Steps: How to Keep Preparing

You've completed the scrimmages—great work. But your prep doesn't stop here. There are different ways to continue sharpening your ACT skills. Below, you'll find the Focus Areas for Math. It highlights which topics tend to show up most on the test. Pair that with one of three strategies to guide your study plan:

1. **Double Down on Your Strengths**
   Keep practicing what you're already strong at. This locks in easy points and helps you move quickly on test day, freeing time for harder questions.

2. **Prioritize High-Frequency Topics**
   Focus on what shows up most often. Even if you're not perfect, being solid here guarantees a higher return for your effort.

3. **Tackle Your Weaknesses**
   If there's an area that consistently slows you down, spend time there. Improving even a little can boost your confidence and raise your score.

No single approach is "best"—the right choice depends on what you think is best (you know yourself better than anyone else) and how much time you have before test day.

Math Focus Areas

The ACT Math section focuses on six content areas: Pre-Algebra, Elementary Algebra, Intermediate Algebra, Coordinate Geometry, Plane Geometry, and Trigonometry. Key topics include linear equations, quadratic equations, functions, geometric formulas, probability, and data analysis, with a trend towards applying these concepts in word problems.

Top Content Areas & Topics

Pre-Algebra (Fractions, Ratios, Decimals, Percentages, Negative Numbers): Basic operations and conversions, understanding parts and wholes, and working with different types of numbers.

Elementary Algebra (Solving Equations, Exponents, Functions): Manipulating algebraic expressions, understanding properties of exponents and roots, and function notation.

Intermediate Algebra (Quadratic Equations, Systems of Equations, Radicals): Factoring, solving complex equations, and working with square roots and other radicals.

Coordinate Geometry (Graphs, Slopes, Midpoints): Plotting points, finding distances and midpoints, interpreting graphs, and understanding slope-intercept form.

Plane Geometry (Shapes, Angles, Circles, Pythagorean Theorem):

Properties of triangles, quadrilaterals, and circles, as well as the Pythagorean theorem for right triangles.

Trigonometry (SOHCAHTOA, Radians/Degrees): Basic trigonometric ratios (sine, cosine, tangent) in right triangles and conversions between radians and degrees. Important Skills and Strategies

Important Skills and Strategies

Word Problems: Translating real-life scenarios into mathematical equations and solving them.

Formulas: Memorizing common formulas for area, volume, slope, and the quadratic formula.

Data Analysis: Interpreting data sets and understanding statistical concepts like the mean.

Picking Numbers/Back solving: Using strategies like picking numbers to simplify problems or working backward from the answer choices.

From here, you can decide what to do: strengthen your strong spots, focus on what the test throws at you most often, or chip away at your weak areas. **The best prep plan is the one that fits your timeline before test day.**

1. Answer: B / Explanation: Subtract 3 from both sides: $2x = 8$. Then divide by 2: $x = 4$.

2. Answer: A / Explanation: Sum of angles in a triangle is 180°. So, 180 - 35 - 65 = 80°.

3. Answer: A / Explanation: Slope = (11 - 3) / (6 - 2) = 8 / 4 = 2.

4. Answer: A / Explanation: Distribute: $3x - 6 + 4x = 7x - 6$.

5. Answer: A / Explanation: $f(2) = 2^2 - 3\times2 + 2 = 4 - 6 + 2 = 0$.

6. Answer: B / Explanation: $3^2 + 4^2 = 9 + 16 = 25$.

7. Answer: A / Explanation: Combine like terms: $5x - 2x = 3x$, so the answer is $3x + 7$.

8. Answer: B / Explanation: Slope-intercept form is $y = mx + b$, so the slope is −2.

9. Answer: C / Explanation: $2(x - 1) = 10 \rightarrow x - 1 = 5 \rightarrow x = 6$.

10. Answer: D / Explanation: Sum of angles = 180°. So, 180 − 110 = 70°.

11. Answer: A / Explanation: Factors of 6 that add to −5 are −2 and −3.

12. Answer: B / Explanation: $f(-3) = 2(-3)^2 = 2\times9 = 18$.

13. Answer: A / Explanation: Midpoint = ((2+6)/2, (3+7)/2) = (4, 5).

14. Answer: B / Explanation: $x(x - 4) = 0 \rightarrow x = 0$ or 4.

15. Answer: A / Explanation: 0.4 × 60 = 24.

16. Answer: C / Explanation: Perimeter = 2(10 + 4) = 28.

17. Answer: A / Explanation: (3×8)/(4×9) = 24/36 = 2/3.

18. Answer: A / Explanation: $x + 2 = 3x - 4 \rightarrow 6 = 2x \rightarrow x = 3$.

19. Answer: A / Explanation: Sum = 30, Average = 30 / 4 = 7.5.

20. Answer: A / Explanation: $y = 2\times5 = 10$.

21. Answer: B / Explanation: 36. 25% of 20 is 5. \$20 - \$5 = \$15

22. Answer: A / Explanation: Area = $\pi r^2 = \pi \times 3^2 = 9\pi$.

23. Answer: A / Explanation: Sorted set: 2, 4, 6, 8, 10 → middle value is 6.

24. Answer: A / Explanation: 39. Area = 0.5 × base × height = 0.5 × 10 × 5 = 25.

25. Answer: B / Explanation: 50 = 10 x height / 2. 50 x 2 = 100. 100 / 10 = 10

26. Answer: D / Explanation: Each term ×2 → 16×2 = 32

27. Answer: B / Explanation: Same base: subtract exponents: X^2

28. Answer: C / Explanation: 43. 2+3 = 5 parts. 20 ÷ 5 = 4. Girls: 3×4 = 12.

29. Answer: A / Explanation: Slope = (7-3)/ (4-2) = 4/2 = 2.

30. Answer: A / Explanation: 4+6 = 10 parts. 20 ÷ 10 = 2. Girls: 2×4 = 8.

31. Answer: A / Explanation: Slope = (7-3)/ (4-2) = 4/2 = 2.

32. Answer: B / Explanation: $10^2 = 100$, so $\log_{10}(100) = 2$.

33. Answer: C / Explanation: Add 5 to both sides: $2x = 14$, then divide: $x = 7$.

34. Answer: A / Explanation: (x -2) and (x-1) are the factors

35. Answer: D /Explanation: Same base: add exponents: 3^(2+3) = 3^5 = 243.

1. Answer: C / Explanation: Servings scale proportionally with juice. $\frac{4cups}{10servings} = \frac{7cups}{??}$ Always cross multiply, then divide so 7 x 10 ÷ 4 = 17.5

2. Answer: B / Explanation: 40 members total, but 5 officers are ineligible → 35 eligible.

Jordan is 1 of those 35 → probability $= \frac{1}{35}$

3. Answer: H / Explanation: City Gym 50+4c. Community Gym 40+6c.

Set equal: 50+4c = 40+6c ⇒10 = 2c ⇒ c =5

4. Answer: J / Explanation: Mean = median (given) and sum is 475. Median $= \frac{475}{5}$ = 95. Sum of the 4 non-median scores = 475−95 =380

5. Answer: H / Explanation: Fine is $25 per second, and the total was $300. 300 ÷ 25 = 12

6. Answer: D / Explanation: 180 ÷ 12 = 15. That is the first answer choice less than 16.

7. Answer: A / Explanation: Legs of the right triangle are 12 and 16 ($a^2+b^2=c^2$). 144 + 256 = 400. $\sqrt{400}$ = 20

Distance from X to Z is half of the hypotenuse $= \frac{hyp}{2}$ = 10

8. Answer: H / Explanation: Let G be Greater and L be Lesser. G + L = 119. L = $\sqrt{G}$+ 9. G + $\sqrt{G}$+ 9 = 119. G + $\sqrt{G}$ **= 110.** Try a perfect square close to 110 (100 or 121) $\sqrt{121}$ = 11. 121 + 11 = 132 not 110. $\sqrt{100}$ = 10. 10 + 100 = 110 ✓ Since G is 100 and L = $\sqrt{100}$ + 9 then L is 19. 100 – 19 = 81.

9. Answer: C / Explanation: Each coin: P(heads) $= \frac{1}{2}$, 4 points for heads. Expected per coin = 4 x $\frac{1}{2}$ = 2.

Three coins → 3×2 =6

10. Answer: B / Explanation: Base area = 12 · 5 =60 cm2. Water rises from 3 to 5.4 ⇒ Δh = 2.4 cm.

Displaced volume = 60 · 2.4 = 144 cm3. Closest choice: 150.

11. Answer: C / Explanation: Area = 18 · 27 = 486 ft^2. 1 yd^2 = 9 ft^2 ⇒ 486 ÷ 9 =54 yd^2

12. Answer: G / Explanation: Emma ate 3 parts. It's 5 parts left ($\frac{5}{8}$) so divide that by 5 and you get $\frac{1}{8}$ each.

13. Answer: G / Explanation: |3x−9|+4=10⇒|3x−9|=6 Split absolute value: 3x −9 = 6 or −(3x −9) = 6

14. Answer: C / Explanation: 5x = 2y −10 ⇒ 2y = 5x+10 ⇒ y= $\frac{5}{2}$x + 5

15. Answer: B / Explanation: 6x = 18 ⇒ x = 3. 3y − 5 = 16 ⇒ 3y = 21⇒ y = 7. Sum = 3+7 = 10

16. Answer: K / Explanation: Odd + even = odd (always).

17. Answer: E / Explanation: Original average 92 ⇒ sum = 4 · 92 = 368. Replace 76 with 100 → new sum = 368 – 76 + 100 = 392. New average = 392/4 = 98.

18. Correct Answer: D / Explanation: Determinant of $\begin{bmatrix} k & 7 \\ 8 & k \end{bmatrix}$ is K x K – 7 x 8 = K^2 – 56. Now set equal to K, per instructions. K^2 −56 = k ⇒ k^2−k−56 = 0. Factor: (k−8)(k+7)=0 ⇒ k= 8 or k = −7.

19. Answer: D / Explanation: Numbers: 30, 60, 70, 82, X, Y. Mode = 30 ⇒ one of is 30. To make median 50, the middle two must average 50 so the other missing variable is 40. We know have 30 30 40 60 70 82 and divide the sum of them (312) by six to get 52.

20. Answer: C / Explanation: Two points (−3, 4) and (3, −5). Slope = $\frac{-5\ -4}{3-(-3)} = \frac{-9}{6} = -\frac{3}{2}$

21. Answer: B / Explanation: Midpoint = 6 + 2 ÷ 2 = 4 and 10 + (−6) ÷ = 2 so (4, 2)

22. Answer: B /Explanation: from 6 to 10 it increased 5 total. 4 divided by 4 is 1.25. Each term differs by 1.25, so subtract each term, starting with the 6th, by 1.25 to get the first four, then add them up.

23. Answer: C / Explanation: for these problems, you can use the arithmetic formula: a☐ = a + (n-1) d where a☐ represents the nth term and a represents a term. If it has a 1 under it, then it's the first term. In this problem, we'll use the fifth term, so it'd be a_5. (d) is always the common difference. In this problem, it's 6. a☐ (the 50th term we're solving for) = 20 + (50 – 5) x 6 = 20 + 45 x 6 = 290

24. Answer: C / Explanation: You can cancel like terms on both sides, so for answer choice "C", cancel out the -5's on both sides and the 3's in the denominator. That leaves you with x > y, which is true.

25. Answer: C / Explanation: AB is 24 and FE is 18. 24-18 is 6. Half(way) of 6 is 3. 18 + 3 is 21. That's how far east you go. FA is 28 and DE is 10. 28 – 10 = 18. Half(way) of 18 is 9.

1. C. absolute value of both is 3. 3 – 3 = 0

2. F. 0.3 x ? = 12. Divide both sides by 0.3 to get ? = 40, then take 5% of 40.

3. D. 5 is given for all sides of the square, and an equilateral triangle's sides are all equal, so all five sides that make up the perimeter equal 25.

4. G. 2x + 9 = 15. Subtract 9 from both sides, then divide both sides by 2 to get x = 3.

5. B. Plug in 2 everywhere there's an "x" in the equation, then solve for x.

6. K. each term is multiplied by 3, and switches from positive to negative each time.

7. C. Factors of 45 that must include five are 9 and 5. 9 + 9 + 5 + 5 = 28.

8. J. Opposite angles are congruent, so C is 35° for both and 35 + 45 is 80. 180 – 80 gives you ∠ BAC.

9. C. Plug in the answer choices for x and add them up. Always start in the middle and determine whether to go up or down if needed.

10 F. Distribute $\frac{1}{2}y$ and then combine like terms.

11. C. You can plug in numbers to test it. When plugging in numbers, use low numbers like 2-5 and avoid 0 and 1.

12. G. Similar triangles have proportional sides (and the same angles), so 7.5-15 means 6-12 (they both doubled). Set up a ratio proportion ($\frac{7.5}{12.5} = \frac{6}{x}$) to solve for the last line, and then add up the numbers for the perimeter.

13. C. The shorter piece is 2, and the total is 5. Set up your ratio proportion ($\frac{2}{5} = \frac{x}{50}$) to solve.

14. K. When you have the same base (d in this instance) and you're dividing exponents, you subtract the exponents. 4-6 is d^{-2}, and it can also be written as $\frac{1}{d^2}$

15. A. Use the formula ($2\pi r^2 + 2\pi rh$). r = 5 and h = 10.

16. F. Sine: $\sin\theta = \frac{opposite}{hypotenuse}$. The hypotenuse is given, and to solve for the missing side, use the Pythagorean theorem: $a^2 + b^2 = c^2$. The unknown side (DE) is 6. 6/10 reduces to 2/3.

17. D. Insert the values for the given formula. 1 + 0.02 = 1.02^3 = 1.061208 x 15,000 = 15,918.12

18. H. 50/200 x 100 = 25%.

19. D. The total is 1,000, and the sample was 200. 1,000/200 = 5. 5 x 80 = 400.

20. G. $\frac{40}{200} = \frac{x}{360}$ to solve for the angle.

21. C. When you multiply the square root of a number by the square root of that same number, you get the number without the square root. Example: $\sqrt{3} \times \sqrt{3} = \sqrt{9} = 3$.

22. F. Use the slope formula: $(y_2 - y_1) / (x_2 - x_1)$

23. E. In three days, 5 books are read, so it's between D and E. If they read for three more days, the total needs to be 10. Only E gives you that answer.

24. K. Base area of tank = $50 \times 30 = 1500$ cm^2. Rise in water level = 0.35 cm. Multiply base area × rise in height = $1500 \times 0.35 = 525$ cm^3. By displacement, the volume of water displaced equals the volume of the object.

25. D. Originally, it's 1/8 f(ull). Add 20 c(ups) 1/8 F + 20C = ¾ F (ull). 20C = 3/4 ÷ 1/8 = 5/8. 20C = 5/8F. Divide both sides by 5/8 to get 32C.

Based on the topics covered in the ACT math section—including algebra, geometry, and trigonometry—the following are top jobs that require these mathematical skills

10 jobs that require ACT Math skills

1. **Actuary:** Actuaries use mathematics, statistics, and financial theory to analyze risk and uncertainty. They primarily work in the insurance industry, helping to design policies and pension plans. (Statistics, probability, algebra.)

2. **Data Scientist:** Data scientists use strong mathematical and statistical skills to analyze large, complex datasets, extracting meaningful insights that help companies make strategic decisions. (Statistics, probability, algebra, functions.)

3. **Financial Analyst:** Financial analysts assess the performance of stocks, bonds, and other investments. They use mathematical modeling to predict financial outcomes and provide guidance to clients and companies. (Algebra, statistics, probability)

4. **Engineer:** Engineers across various disciplines rely heavily on ACT-level math, often using it as a foundation for more complex calculations.

- **Civil Engineer:** Designs and supervises the construction of infrastructure like roads, bridges, and dams. This requires skills in geometry and algebra to calculate measurements and structural integrity.
- **Mechanical Engineer:** Creates and improves mechanical devices and systems. They apply algebra and geometry to analyze mechanics, stress, and heat transfer.
- **Software Engineer:** Designs and develops software programs, which requires a strong grasp of algebra and algorithms. (Geometry, algebra, trigonometry.)

5. **Statistician:** Statisticians collect, interpret, and present numerical data to identify trends and solve real-world problems. They are employed in many fields, including business, healthcare, and government. (Statistics, probability, algebra, functions.)

6. **Market Research Analyst:** These analysts help companies understand consumer behavior and market trends by designing surveys, collecting data, and analyzing

statistics. Their mathematical findings inform marketing and business strategies. (Statistics, probability, algebra.)

7. **Economist:** Economists study the production, distribution, and consumption of goods and services. They use mathematical models to analyze economic issues, forecast economic trends, and provide guidance to organizations and governments. (Statistics, algebra, probability.)

8. **Accountant:** Accountants manage financial records for individuals and companies, preparing financial reports and ensuring legal compliance. The work requires a strong foundation in algebra for managing budgets and forecasting. (Algebra, pre-algebra.)

9. **Architect:** Architects design buildings and other structures. They use geometry and trigonometry to create precise designs, ensuring structural soundness and safety. (Geometry, trigonometry, algebra.)

10. **Math Teacher:** A math teacher, particularly at the high school level, directly applies the full range of ACT-level math skills every day. They develop lesson plans for subjects like algebra, geometry, and trigonometry. (All ACT math topics.)

READING

Chapter 7: Reading Stretch

Passage 1: Maria carefully studied the patterns of migration among monarch butterflies. She observed that their journey spanned thousands of miles, crossing multiple countries and climates. Scientists believe these migrations are triggered by seasonal changes, but Maria suspected other factors might also influence their routes, such as wind patterns and habitat destruction.

What is the main idea of the passage?

A. Maria believes seasonal changes are the only reason for butterfly migration.

B. Monarch butterfly migration spans great distances and may be influenced by multiple factors.

C. Wind patterns are the main influence on butterfly migration.

D. Scientists disagree with Maria's theories on butterfly migration.

Passage 2: Over the past decade, urban gardens have grown in popularity. From rooftops to abandoned lots, city dwellers are finding creative ways to cultivate fruits and vegetables. Advocates claim these gardens not only provide fresh produce but also foster a sense of community and sustainability.

What is the main idea of the passage?

A. Urban gardening has become more common and has benefits like fresh food, community building, and sustainability.

B. Abandoned lots are the most popular location for new urban gardens.

C. Urban gardens are primarily for growing fruit, not vegetables.

D. Urban gardening reduces the need for grocery stores.

Passage 3: Despite the team's defeat, Coach Ramirez spoke with pride. She highlighted their growth, teamwork, and determination throughout the season. Her words carried hope, urging the players to see this as a stepping stone rather than a setback.

What best describes the author's tone?

A. Neutral and detached

B. Disappointed and frustrated

C. Encouraging and optimistic

D. Sarcastic and mocking

Passage 4: A recent editorial criticized the city's failure to address rising pollution levels. The author expressed frustration over the lack of government action, warning that delaying solutions would lead to irreversible consequences.

What best describes the author's tone?

A. Supportive and calm

B. Critical and urgent

C. Indifferent and distant

D. Humorous and casual

Passage 5: During a museum tour, visitors spent the most time examining the fossil exhibits. Attendance records showed that over 70% of guests selected the guided paleontology experience instead of modern art or astronomy sections.

What can reasonably be concluded from the passage?

A. Fossil exhibits are the most popular among visitors

B. Visitors are less interested in science than in art

C. Paleontology is mandatory for all guided tours

D. Astronomy exhibits are the least visited

Passage 6: An online bookstore's sales data revealed that mystery novels consistently outsold every other genre over the past three years. The company recently expanded its mystery selection and added special promotions.

What can reasonably be concluded from the passage?

A. Readers prefer shorter books

B. Expanding the mystery selection aligns with customer preferences

C. Mystery promotions are more expensive than other genres

D. Sales of non-fiction books are declining

Chapter 8: Reading warm-up and drills

As we head into the Reading warm-up phase, think of this as your "film study" session—the place where we break down plays, analyze patterns, and sharpen instincts before game time. The ACT Reading section is less about memorization and more about strategy. You'll face four passages—prose fiction, social science, humanities, and natural science—and the goal isn't just to read but to **read with purpose**.

The ACT Reading section is all about patterns. Each passage has its own "personality," and the test-makers love to disguise traps in subtle ways. They'll test your ability to grasp the main idea, interpret the author's tone, make inferences, and draw logical conclusions while keeping an eye on the clock.

Here's the good news: once you understand how these passages are structured and how questions are framed, the Reading section becomes less of a mystery. Preparation helps you spot common distractors, zero in on evidence, and move confidently through each passage.

Think of every drill in this section as a rep toward becoming a stronger, faster, and more strategic reader. By the time you hit the scrimmage and eventually test day, you'll have the pacing, accuracy, and confidence to handle whatever passages come your way.

Now, let's dive into the drills, sharpen your comprehension skills, and build your reading game plan.

SCOUTING
REPORT
KNOW THE
PLAY.
MASTER THE GAME.
MOORE SUBSTANCE
TEST PREP

ACT Reading Strategies & Tips

1. Skim Before You Dive

- Don't start by reading the questions first—it will slow you down.
- Spend the first 3-4 mins skimming the passage to get a sense of the structure, tone, and where key information is located.
- Think of it like creating a mental map of the passage—**location is everything**.

---

2. Trust the Author, Not the Facts

- This isn't a history or science test. You're tested on what the author wrote, not reality.
- If the passage says, *"the sky is green,"* then it's green—don't argue with it.
- Stick to the author's version of facts. ACT loves to trap students who assume they "know better."

---

3. Use Line References Wisely

- When a question refers to specific line numbers, go directly to those lines.
- The answer will only come from those lines—not before, not after.
- ACT loves to put tempting, wrong choices right outside the referenced lines to throw you off.

---

4. Eliminate Wrong Answers Aggressively

- If an answer choice isn't in the passage, it's wrong—no matter how good it sounds.
- Narrow it down to one choice by proving three are wrong.
- Remember: your job isn't to find the "best-sounding" answer—it's to find the correct one.

5. Stay Locked In

- It's normal to zone out during longer passages, but drifting costs points.
- Remind yourself: every correct answer is worth **big bucks** in scholarship opportunities.
- Stay focused, stay sharp, and keep pushing forward.

6. Manage Your Time Like a Pro

- You'll have 40 minutes for 36 questions—less than 9 minutes per passage.
- Don't spend too much time on any single question or passage.
- All passages are worth the same points, so don't get stuck thinking the early ones are more important—the last passage could be your easiest.

7. Treat It Like an Open-Book Test

- Every answer is somewhere in the passage.
- But remember—it's an open book with a timer. Speed and accuracy are key.

8. Build Your Reading Muscles

- The ACT Reading section tests you at a college level.
- To prepare:
  - Read 10–15 minutes daily—articles, essays, and stories.
  - Grow your vocabulary steadily to make challenging passages easier to handle.

9. Practice Under Pressure

- The more timed drills you do, the better you'll perform.
- Simulate real test conditions:

- Set a timer (67 secs per question) when you're testing and don't go over.
- Review mistakes immediately to lock in learning.

---

10. Annotate as You Skim

- Don't just skim — **engage with the text**.
- Quickly circle names, underline key phrases, and mark transitions ("however," "for example," "in contrast").
- Light annotations act as a **roadmap**, helping you locate evidence faster when answering questions.
- The goal isn't to rewrite the passage — it's to leave **breadcrumbs** so you can retrace your steps quickly under time pressure.

Let's start using these Reading strategies. Give the same effort in practice as you would the real thing: muscle memory is real.

Passage A: Unlike their contemporary, James Audubon, whose artistic flair dominated early American ornithology, Wilson adopted a methodical and scientific approach. His field observations and commitment to accuracy in depiction made him a forerunner in modern scientific illustration. Wilson was not only an artist and notable ornithologist, but also a schoolteacher and poet. His poetic sensibilities lent a narrative grace to his journal entries, while his scientific rigor grounded his work in credibility. Though not as flamboyant or widely celebrated in his time, Wilson's influence rippled through the scientific community and inspired the likes of Charles Lucien Bonaparte and later John James Audubon himself.

1. What is the central focus of the passage?

A. Wilson's lack of recognition during his lifetime compared to James Audubon

B. Wilson's contributions to scientific illustration and ornithology

C. The evolution of modern poetry and journal writing in scientific studies

D. The rivalry between Wilson and other prominent ornithologists

2. What is the author's attitude toward Wilson?

A. Critical of his lack of fame

B. Respectful of his contributions

C. Amused by his poetic tendencies

D. Indifferent to his work

3. The word 'flamboyant' most nearly means:

A. Subtle

B. Showy

C. Efficient

D. Quiet

4. Which statement is supported by the passage?

A. Wilson had a more lasting legacy than Audubon.

B. Wilson's illustrations were more poetic than accurate.

C. Wilson made significant contributions to ornithology.

D. Wilson and Audubon frequently collaborated.

5. The mention of Charles Lucien Bonaparte serves to:

A. Show the political relevance of Wilson's work.

B. Undermine Audubon's contributions.

C. Highlight Wilson's influence on future scientists.

D. Introduce a new ornithological theory.

Passage B: In the early 20th century, urban planning emerged as a formal profession. Cities in Europe and North America were rapidly expanding due to industrialization. This growth created overcrowded housing, traffic congestion, and poor sanitation. Urban planners sought to design cities with better living conditions, focusing on zoning, green spaces, and public transportation. In the United States, the City Beautiful movement aimed to introduce beautification and monumental grandeur into urban spaces. Planners believed that attractive environments would inspire civic pride and moral virtue. Although some criticized the movement for prioritizing aesthetics over social reform, it laid the groundwork for modern urban planning principles still used today.

1. What was a major reason urban planning emerged as a profession in the early 20th century?

A. Decline in rural populations

B. Need for improved urban living conditions

C. Advances in technology

D. Expansion of the suburbs

2. What was one feature of the City Beautiful movement?

A. Construction of skyscrapers

B. Reduction of city population

C. Emphasis on beautification and grandeur

D. Use of modern art in city layouts

3. What challenge did urban planners NOT focus on according to the passage?

A. Traffic congestion

B. Poor sanitation

C. Overcrowded housing

D. Agricultural reform

4. What belief motivated planners involved in the City Beautiful movement?

A. Urban beautification could lead to civic pride and moral virtue

B. Technology would solve all urban problems

C. Cities should be designed for cars

D. Parks were unnecessary luxuries

5. What was one criticism of the City Beautiful movement?

A. It ignored the need for public transportation

B. It prioritized aesthetics over social reform

C. It relied too much on foreign influence

D. It caused further overcrowding

Passage C: In the early 20th century, urban planning in America underwent a significant transformation. Driven by rapid industrialization and a growing population, cities began to shift from organic, unregulated growth to more structured and intentional designs. One of the key figures in this movement was Frederick Law Olmsted Jr., son of the famed Central Park designer. Olmsted Jr. helped establish city planning as a legitimate profession, emphasizing the integration of green spaces, functional zoning, and human-centric design.

The City Beautiful movement, which emerged in the 1890s and influenced planning into the 1920s, aimed to introduce beauty and order to chaotic urban environments. Proponents believed that attractive surroundings would inspire civic virtue and reduce social unrest. This philosophy resulted in grand boulevards, monumental buildings, and the expansion of public parks.

However, critics argued that City Beautiful projects often prioritized aesthetics over practicality and displaced marginalized communities in the name of progress. As urban planning evolved, a balance between beauty, equity, and functionality became the gold standard, setting the stage for the more inclusive planning efforts of the late 20th and early 21st centuries.

1. According to the passage, what was a key motivation for changing urban planning strategies in early 20th-century America?

A. A desire to preserve historic landmarks

B. The need to accommodate a rising population and industrial growth

C. Pressure from rural communities

D. The influence of European architectural styles

2. The City Beautiful movement believed that aesthetically pleasing urban areas would:

A. Encourage private investment

B. Improve public transportation

C. Foster civic responsibility and reduce conflict

D. Expand suburban housing developments

3. What criticism of the City Beautiful movement is mentioned in the passage?

A. It failed to use modern construction methods

B. It ignored traffic congestion issues

C. It focused too much on green technology

D. It often displaced underrepresented populations

4. The author most likely views Frederick Law Olmsted Jr. as:

A. A political reformer with radical ideas

B. A pioneer who helped formalize urban planning

C. An artist who favored form over function

D. A critic of traditional architecture

5. Which of the following best summarizes the evolution of urban planning discussed in the passage?

A. From rural design to suburban expansion

B. From unregulated growth to balanced, inclusive planning

C. From modern architecture to classic revival styles

D. From individual projects to regional zoning enforcement

Passage D: In the early 20th century, the art world experienced a significant shift as Modernism began to take root. Traditional artistic conventions were challenged by new movements such as Cubism, Futurism, and Dadaism. These movements sought to break away from realistic representation and instead embraced abstraction, fragmentation, and conceptual exploration. One notable artist, Wassily Kandinsky, believed that color and form could evoke emotion independently of visual reference to the real world. Meanwhile, the Dada movement, with figures like Marcel Duchamp, rejected the very premise of conventional art, presenting ready-made objects as artworks to provoke, question, and destabilize accepted norms.

1. The primary purpose of the passage is to:
A. analyze the psychological motivations behind Modernist artists' work.
B. describe the emergence and characteristics of Modernist movements.
C. compare traditional and modern art techniques in-depth.
D. argue that abstract art is superior to realistic art.

2. The author uses Wassily Kandinsky to illustrate:
A. the resistance some artists had to abstract art.
B. how color and form were used to replicate reality.
C. how abstract elements could independently convey emotion.
D. the competition between Dadaism and Cubism.

3. The word 'fragmentation' most nearly refers to:
A. the splitting of movements into different time periods.
B. the division of subjects into broken or disjointed parts.
C. a systematic destruction of the art world.
D. the classification of artists by style.

4. Which of the following best describes the Dada movement's goal?

A. To elevate landscape painting to the highest form of art.

B. To present unusual objects in a humorous or shocking context.

C. To develop precise representations of urban life.

D. To establish strict rules for modern artistic expression.

5. Based on the passage, how did Modernism differ from traditional art?

A. Modernism prioritized realism over symbolism.

B. Modernism focused on moral storytelling.

C. Modernism embraced unconventional ideas and forms.

D. Modernism rejected all color and form in art.

Passage E: In the early 20th century, American painter Georgia O'Keeffe defied conventional artistic norms with her distinct and often abstract floral paintings. While many critics tried to interpret her work as overtly symbolic, O'Keeffe maintained that her intent was purely to depict the beauty she saw in nature. Her larger-than-life flowers, rendered in vibrant colors and sweeping lines, invited viewers to observe the minute details they might otherwise overlook. O'Keeffe's insistence on controlling the interpretation of her own work positioned her as a pioneering voice in the modernist movement and helped pave the way for future generations of female artists.

1. What is the central focus of the passage?

A. The rise of abstract art in the early 20th century

B. Georgia O'Keeffe's impact on modernist painting and her artistic intent

C. The symbolic interpretations of floral paintings in modern art

D. A comparison between Georgia O'Keeffe and her contemporaries

2. According to the passage, how did O'Keeffe respond to critics' interpretations of her work?

A. She appreciated their interest but encouraged diverse views.

B. She agreed that her work was deeply symbolic.

C. She rejected symbolic interpretations and clarified her intent.

D. She remained silent and let the work speak for itself.

3. The author of the passage most likely views O'Keeffe as:

A. a misunderstood traditionalist.

B. a reluctant artist.

C. a groundbreaking and influential figure.

D. a critic of modernism.

4. What literary device is used in describing O'Keeffe's flowers as "larger-than-life"?

A. Simile

B. Hyperbole

C. Irony

D. Allusion

5. What does the passage suggest about O'Keeffe's legacy?

A. She was mostly forgotten until the late 20th century.

B. Her work was overshadowed by other modernist painters.

C. She set a precedent for female artists to define their own narratives.

D. Her influence was confined to floral still lifes.

Passage F: **(The Last First Step)** Amara's heart raced as she stood at the edge of the gymnasium floor. The echo of squeaky sneakers and bouncing basketballs was a comforting chaos she had grown up hearing, but today, it all felt foreign. This was her first tryout for the varsity team—a step she'd been preparing for since she first dribbled a ball in kindergarten.

She looked down at her shoes—new ones, crisp and glaringly white—and thought of how much her older brother teased her for being superstitious about sneakers. He used to say the shoes didn't make the player, but she wasn't so sure. These shoes had been waiting in their box for this very moment, and she needed every drop of confidence they could give her.

Coach Valdez blew the whistle, and the first drill began. Amara took off, weaving between cones, feeling her nerves settle into the rhythm of motion. She caught a glance from another player—Tasha—who was fast, focused, and clearly here to win. The competition was stiff, and Amara knew it. Still, something sparked inside her—not fear, but resolve.

Later, during the scrimmage, Amara made a pass that surprised even her: a clean, no-look assist that brought Coach Valdez's eyebrows up and earned a quick nod. It wasn't flashy, but it was smart—decisive. She played the rest of the game like she'd already made the team, not arrogantly, but with intention.

When the whistle finally blew to end the session, Amara stood tall. No matter what happened next, she had taken her first step. And maybe—just maybe—that was enough.

1. What is the main focus of this passage?

A. The challenges Amara faces balancing school and sports

B. Amara's relationship with her brother

C. A pivotal moment during Amara's varsity basketball tryout

D. A rivalry between Amara and another player

2. Which of the following best describes Amara's mindset by the end of the tryout?

A. She is filled with self-doubt about her performance

B. She regrets not practicing more beforehand

C. She accepts the outcome and feels confident in her effort

D. She becomes nervous after seeing Coach Valdez's reaction

3. The description of Amara's shoes serves what purpose in the passage?

A. To show her passion for shopping

B. To highlight her attention to appearance

C. To symbolize her belief in preparation and confidence

D. To suggest she cares more about image than performance

4. What can be inferred about Tasha, the player Amara notices during the drill?

A. Tasha is Amara's closest friend

B. Tasha is likely a returning varsity player

C. Tasha distracts Amara and causes her to perform poorly

D. Tasha motivates Amara to perform her best

5. Based on the passage, how does Coach Valdez likely view Amara after the tryout?

A. He is disappointed in her lack of teamwork

B. He is indifferent to her performance

C. He is impressed by her court vision and smart play

D. He believes she needs more training before joining varsity

Passage G: **(The Lucky Ones)** When Jordan moved to the city for college, everything felt overwhelming—the trains, the crowds, even the endless rows of coffee shops that all somehow felt the same. Back in her small town, there were no subway lines or bodegas open at 2 a.m. There was just one gas station, three stoplights, and a high school where everyone knew your name and who your older siblings had dated.

But Jordan wasn't in Pineville anymore. She was here—at a university with students who wore vintage coats and spoke four languages and casually referenced philosophers over dinner. She spent the first month just trying not to get lost between classes and deciphering syllabi that read like legal contracts.

Things started to shift when she met Zoe in a writing seminar. Zoe was from Chicago, confident in a way Jordan admired—not flashy, just certain. When Jordan confessed that she sometimes missed home, Zoe didn't laugh. She nodded and said, "Me too. City's loud."

They became friends, sitting in the library for hours, challenging each other to write better, clearer, sharper. Zoe's friendship was like a light—quiet, steady, always there.

By spring, Jordan found her rhythm. She still missed Pineville, but she didn't feel lost anymore. She knew which train to take and had a favorite coffee shop where they learned her name. One day she realized she wasn't pretending to be a city girl anymore—she simply was.

1. What is the main focus of the passage?

A. Jordan's academic struggles

B. A small-town girl adjusting to city life

C. A rivalry between classmates

D. A student's dislike for college

2. How does Jordan feel about the city at first?

A. Excited and energetic

B. Confused and overwhelmed

C. Disappointed and bored

D. Confident and at ease

3. What role does Zoe play in the passage?

A. Jordan's academic rival

B. A casual classmate

C. A supportive friend

D. A tutor assigned by the university

4. What is implied about Jordan's transformation by spring?

A. She wishes to return home

B. She no longer values her small-town background

C. She decides to transfer to another school

D. She is more confident and feels a sense of belonging

5. Why does the author describe Pineville?

A. To contrast it with city life

B. To mock small towns

C. To suggest Jordan regrets leaving

D. To highlight Jordan's family history

Passage H: In a world racing toward innovation, the quiet dignity of historic buildings often goes unnoticed. Yet these structures, with their weathered facades and timeworn craftsmanship, tell stories that connect us to our roots. As a historical preservationist, I believe that restoring and repurposing our architectural past is not merely about nostalgia—it's about cultural continuity and sustainable living.

A century-old courthouse may no longer serve its original function, but its pillars still echo the civic ideals upon which a community was founded. The intricate woodwork in a Victorian home, painstakingly carved by hand, represents a level of artistry that mass production has largely forgotten. And while some see dilapidation, I see opportunity—to reinvigorate neighborhoods, foster tourism, and reduce environmental waste by adapting existing structures instead of demolishing them.

Preservation isn't resistance to progress; it's progress with perspective. It reminds us that the future doesn't require erasing the past. In fact, honoring our history can inspire better design, stronger identity, and a more rooted sense of place.

1. What is the central argument made by the author?

A. Modern buildings are superior to older architecture.

B. Historic buildings are obstacles to urban progress.

C. Preserving historical buildings is essential for cultural and sustainable reasons.

D. Demolishing old buildings is better for the environment.

2. What does the author most likely mean by 'progress with perspective'?

A. Only modern buildings should be preserved.

B. Society should move forward while respecting and learning from the past.

C. Historical buildings have no relevance in modern times.

D. Preservation halts all forms of progress.

3. According to the author, what is one practical benefit of preserving old buildings?

A. They can be converted into amusement parks.

B. They prevent the need for environmental legislation.

C. They speed up city development.

D. They help reduce environmental waste.

4. The author's tone can best be described as—

A. Sarcastic and critical

B. Objective and detached

C. Passionate and persuasive

D. Indifferent and formal

5. Which detail from the passage best supports the idea that historic buildings hold artistic value?

A. 'Weathered facades and timeworn craftsmanship'

B. 'Mass production has largely forgotten'

C. 'Reinvigorate neighborhoods, foster tourism'

D. 'The future doesn't require erasing the past'

Passage I: (Paired Passage)

Passage A In 1893, the Chicago World's Fair was a beacon of innovation and design. Architect Sophia Hayden, one of the few women at the time to receive a degree in architecture, designed the Woman's Building for the exposition. Her neoclassical structure showcased art, literature, and inventions by women from around the globe. While her contributions were significant, Hayden's career was stifled by the era's gender norms, and she eventually left architecture altogether.

Passage B During the 1930s, Zora Neale Hurston emerged as a powerful voice in American literature. An anthropologist and novelist, Hurston's work chronicled African American culture and folklore. Her 1937 novel, Their Eyes Were Watching God, explored themes of identity and independence through the lens of a Black female protagonist. Though largely unrecognized in her lifetime, Hurston's work gained prominence posthumously, highlighting her pivotal role in the Harlem Renaissance.

1. Both passages primarily highlight women who:

A. excelled despite societal limitations

B. pioneered modern technology

C. focused on educational reform

D. relied heavily on male mentors

2. Based on Passage A, what was one outcome of Hayden's experience at the World's Fair?

A. She was celebrated and immediately became a national figure.

B. She continued designing neoclassical buildings across the U.S.

C. She abandoned her profession due to societal barriers.

D. She traveled abroad to study further.

3. In Passage B, what best describes the initial reception of Hurston's work?

A. It was praised immediately and widely adopted into curricula.

B. It was largely overlooked during her lifetime.

C. It caused significant controversy within her community.

D. It was published anonymously.

4. What is a central theme common to both passages?

A. The pursuit of wealth and fame

B. Overcoming barriers in male-dominated fields

C. Developing new scientific theories

D. Religious devotion

5. Which of the following best compares the challenges faced by the two women?

A. Both were supported by institutions.

B. Hayden lacked formal training, while Hurston had significant education.

C. Hurston faced no resistance, while Hayden did.

D. Both faced societal limitations in their respective fields.

Chapter 9: Reading Scrimmage

1. You've stretched. You've warmed up. You've run the drills.

Now it's time to scrimmage. This is where practice meets pressure. You're about to dive into a Reading test built in the spirit of the ACT—same timing, same format, same challenge.

2. Remember—scrimmages aren't about perfection. They're about testing your strategies under game-time conditions. Every mistake here is just "film" for us to review, and every correct answer builds your confidence for the real thing. So stay sharp, stay focused, and approach each passage like a pro.

3. If you're taking both passages without stopping, set your timer for 22 minutes and 30 seconds.

If you want to work on **one passage at a time** (10 questions), set your timer for **11 minutes and 15 seconds** and review your answers immediately after.

4. Both pacing strategies will prepare you for the updated ACT format, giving you confidence and control on test day. Trust your training, rely on your strategies, and don't let the clock rattle you.

5. Make note of any tricky question types or passages so you can revisit them later—online, in tutoring, or with extra drills. Every scrimmage gets you one step closer to your ACT goals.

LITERARY NARRATIVE: This passage was adapted from Hamlin Garland's *A Country Schoolhouse* (public domain).

When I was twelve years old, the annual county fair loomed like a distant mountain I had no map to climb. The fair meant contests—livestock, pies, needlework—and for the first time, an essay competition. The theme was *Perseverance in Daily Life*, and the moment I saw the notice tacked to the library wall, my chest tightened with both dread and a strange sort of hunger.

At school, the other students laughed when I told them I was entering. Writing was not the pursuit of someone from my corner of the world, where strong hands for the plow or steady arms for the mill meant more than clever words on a page. My teacher, Miss Walker, simply nodded and told me, "Then write as if your own breath depends upon it." I carried those words with me, heavy as a secret.

In the evenings, while the rest of my family tended chores, I hunched by the oil lamp at our worn oak table. My little sister teased me, saying I stared at the page so hard I'd burn holes clean through it. Truthfully, most nights I did nothing but scratch out false starts, entire paragraphs that fell limp on the paper. Still, I began again each morning, folding fresh courage into every draft.

By late August, my essay stood at three pages—ragged, ink-blotted, but mine. On the day of the fair, I walked the dusty road alone, my essay tucked into the pocket of my best trousers. The exhibition hall smelled of sawdust and roasted peanuts. I remember my hands trembling as I handed the folded sheets to the woman at the registration desk.

When the winners were announced that evening, I did not hear my name among the honored. Instead, the prize went to a boy from town, whose essay had the polish of someone well-schooled in such contests. My throat tightened, but I forced myself to clap with the others.

As I left the fairgrounds, the night air cool against my cheeks, Miss Walker stopped me. She pressed the essay, now marked with judges' notes, back into my hand. "Winning isn't always the measure," she said. "The measure is that you wrote, and you finished."

I walked home beneath the stars, the paper crumpled in my fist, and for the first time I understood that achievement was not only in applause or ribbons. It was in the quiet defiance of trying when no one believed you could.

1. The main purpose of the passage is to:
A. emphasize the narrator's resentment toward losing a contest.
B. illustrate how perseverance can create personal growth even without recognition.
C. describe the rules and traditions of county fair competitions.
D. criticize the unfair treatment of rural children in academic contests.

2. In line 8, when the narrator recalls Miss Walker saying, "Then write as if your own breath depends upon it," this statement serves primarily to:
F. highlight her casual indifference to the narrator's attempt.
G. suggest that writing would be a struggle the narrator could not win.
H. inspire the narrator to approach the essay with deep seriousness.
J. warn the narrator against entering a contest he could not succeed in.

3. The narrator's family's reaction to his writing is described as:
A. skeptical and mocking.
B. supportive but cautious.
C. indifferent, allowing him to pursue it on his own.
D. hostile, actively discouraging him.

4. The phrase "folding fresh courage into every draft" most nearly conveys the narrator's:
F. habit of revising the essay despite frequent discouragement.
G. frustration with starting over repeatedly.

H. perfectionist attitude toward the writing style.

J. belief that writing was a futile exercise.

5. According to the passage, how did the narrator respond to not winning the contest?

A. He was deeply bitter and refused to attend future fairs.

B. He was disappointed but managed to show outward composure.

C. He was relieved because it ended the pressure of competing.

D. He was indifferent since he had expected the loss.

6. In lines 24-26, the narrator says, "I understood that achievement was not only in applause or ribbons. It was in the quiet defiance of trying when no one believed you could." This realization represents:

F. the narrator's recognition of perseverance as a form of victory.

G. his renewed determination to win future contests.

H. his rejection of Miss Walker's earlier encouragement.

J. his regret for ever having entered the contest.

7. The imagery used in describing the exhibition hall, "smelled of sawdust and roasted peanuts," serves to:

A. emphasize the narrator's anxiety in an unfamiliar setting.

B. create a vivid sense of place for the reader.

C. suggest the narrator's disdain for rural traditions.

D. symbolize the bitterness of competition.

8. Which of the following best describes the narrator's attitude by the end of the passage?

F. Bitter resignation.

G. Quiet acceptance and growth.

H. Eager anticipation for future contests.

J. Amused detachment.

9. In the passage, which contrast is most strongly highlighted?

A. The boy from the town's polished essay versus the narrator's ragged draft.

B. The narrator's enthusiasm for writing versus his teacher's support.

C. The narrator's family's constant indifference versus his teacher's encouragement.

D. The prizes of ribbons and applause versus the narrator's lack of interest in them.

10. The narrator's journey in the passage most closely reflects which broader theme?

F. Individual achievement often goes unrecognized by others.

G. Failure in youth prevents later success.

H. Community support is always necessary for achievement.

J. True accomplishment lies in personal perseverance rather than external reward.

HUMANITIES: This passage was adapted from W.E.B. Du Bois's *The Souls of Black Folk* (1903)

I was a boy when I first felt the weight of a strange question pressing against my chest: *"Why is the world divided so?"* The question came not from books, but from the eyes of those around me, from the way doors opened for some and remained closed for others.

In the little schoolhouse, I sat among children who were, in every sense but one, my equals. We played the same games at recess; we puzzled over the same arithmetic at our desks. Yet when a visiting stranger entered, his glance swept past me as though I were invisible, settling instead on the lighter faces of my classmates.

This early experience left me restless. I began to wonder how a nation so proud of its liberty could carry, hidden beneath its flag, a second set of rules. One for those who belonged, and another for those who did not.

As I grew older, I found words for that feeling: *a veil.* It was as if I lived behind a thin but impenetrable curtain. Through it I could see the life of others—opportunity, ease, belonging—but I could not pass through.

I was determined, however, not to let this veil define me entirely. Books became my weapons, education my rebellion. Each page I read pulled me a little closer toward the world beyond the curtain. Still, the veil never disappeared; it only shifted, sometimes closer, sometimes farther away.

Yet within this struggle, there was also a gift: a second sight. To be forced to see the world both as it is and as it might be, to measure truth against possibility, is to gain a kind of clarity. That clarity carried me through my youth, and it remains with me still.

11. The main purpose of the passage is to:

A. describe how education was valued equally by all children.

B. highlight the narrator's frustration with academic failure.

C. illustrate the narrator's struggle to reconcile inequality with hope.

D. argue that social divisions cannot be overcome.

12. In lines 5–7, the narrator's observation of the visiting stranger most directly reveals:

F. the narrator's growing awareness of social exclusion.

G. the narrator's disinterest in academic work.

H. the narrator's amusement at being ignored.

J. the narrator's preference for solitude.

13. What does the "veil" most nearly symbolize?

A. A literal barrier between classrooms.

B. The narrator's sense of separation due to societal norms.

C. The difference between adults and students.

D. The presence of books in daily life.

14. According to the passage, how did the narrator respond to the restrictions imposed by the "veil"?

F. By giving up on pursuing education.

G. By using education as a form of resistance.

H. By ignoring inequality altogether.

J. By demanding immediate recognition from society.

15. The phrase "books became my weapons" most strongly suggests that the narrator:

A. treated reading as a pastime rather than a serious pursuit.

B. saw education as a tool to challenge injustice.

C. believed books were dangerous to personal safety.

D. feared using knowledge to confront inequality.

16. In the context of the passage, the narrator's "second sight" refers to:

F. the ability to memorize books quickly.

G. the skill of predicting his classmates' actions.

H. the gift of perceiving both the world's flaws and its potential.

J. the physical improvement of his eyesight.

17. The imagery of a "curtain" contributes primarily to the passage's:

A. sense of confinement and separation.

B. lighthearted tone about childhood play.

C. emphasis on academic achievement.

D. humor in describing daily life.

18. Which of the following best describes the narrator's attitude toward the "veil" by the end of the passage?

F. Resentful but accepting that it will never lift.

G. Determined to destroy it immediately.

H. Resigned to its permanence, with no hope.

J. Acknowledging its reality but using it to gain clarity.

19. What contrast is most strongly developed in the passage?

A. The narrator's ability to join in childhood play versus his recognition of deeper social exclusion.

B. The narrator's dislike of reading versus his love of arithmetic.

C. The narrator's teacher's encouragement versus his classmates' indifference.

D. The narrator's interest in liberty versus his respect for rules.

20. The broader theme most clearly illustrated by the passage is:

F. Education provides both a refuge and a means of resistance in the face of inequality.

G. Childhood memories are often unreliable accounts of truth.

H. Nations that value liberty never create barriers to progress.

J. Social divisions can only be overcome by avoiding them.

Congratulations—you just powered through the ACT Reading scrimmage! That's no small accomplishment. You faced dense passages, subtle answer choices, and the pressure of the clock, and you came out stronger on the other side.

Here's what this means for you:

- **You built endurance.** Reading four passages back-to-back under timed conditions isn't easy. Now you know what it feels like—and you'll be ready to manage your focus when it counts.
- **You tested your strategy.** From spotting main ideas to analyzing tone and drawing conclusions, you've seen how your approach works when the clock is running—and where you can tighten it up.
- **You practiced under pressure.** The ACT is as much about pacing as it is about comprehension. Every timed rep here builds confidence, speed, and accuracy.

Take a moment to celebrate this step forward. Every question you've tackled, every passage you've broken down, is one more layer of preparation for test day. Keep reviewing your strategies, sharpen your reading skills, add to your vocabulary daily, and carry this momentum into the next section—Science.

Bottom line: you didn't just finish the Reading scrimmage—you leveled up your game. On test day, you'll know how to stay calm, stay confident, and stay in control.

Next Steps: How to Keep Preparing

You've completed the scrimmages—great work. But your prep doesn't stop here. There are different ways to continue sharpening your ACT skills. Below, you'll find the Focus Areas for Reading. It highlights which topics tend to show up the most on the test. Pair that with one of three strategies to guide your study plan:

1. **Double Down on Your Strengths**
   Keep practicing what you're already strong at. This locks in easy points and helps you move quickly on test day, freeing time for harder questions.

2. **Prioritize High-Frequency Topics**
   Focus on what shows up most often. Even if you're not perfect, being solid here guarantees a higher return for your effort.

3. **Tackle Your Weaknesses**
   If there's an area that consistently slows you down, spend time there. Improving even a little can boost your confidence and raise your score.

No single approach is "best"—the right choice depends on what you think is best (you know yourself better than anyone else) and how much time you have before test day.

Reading Focus Areas

The top areas tested on the ACT Reading test are Key Ideas and Details, Craft and Structure, and Integration of Knowledge and Ideas. You'll also need to be familiar with the four predictable passage types: Prose Fiction, Social Science, Humanities, and Natural Science

Content Area Breakdown

ACT, Inc. divides the content of the Reading test into three main categories:

- **Key Ideas and Details (55-60%)**:

This involves understanding the central ideas of a passage, summarizing information, and understanding relationships between parts of the text.

- **Craft and Structure (25-30%)**:

These questions focus on the author's choices, such as word meaning, text structure, and the author's overall purpose and perspective.

- **Integration of Knowledge and Ideas (13-23%)**:

This tests your ability to connect ideas within a passage or between different related texts, including questions about any included graphics.

Passage Types

The test always presents passages in the same order:

1. **Prose Fiction/Literary Narrative**: A story or excerpt from a novel.
2. **Social Science**: Topics like psychology, sociology, history, or geography.
3. **Humanities**: Passages about art, literature, music, or philosophy.
4. **Natural Science**: Information on biology, chemistry, physics, or earth sciences.

Extra Tips for Success

- **Understand the Passage Types**:

Familiarize yourself with the content and question types typical for each genre.

- **Target Your Strengths**:

Since the passages are always in a specific order, you can strategically start with the passage type you're most comfortable with to maximize your score.

- **Know the Question Traps**:

Be aware of common incorrect answers, such as statements that are out of scope, distorted, or too extreme.

From here, you can decide what to do: strengthen your strong spots, focus on what the test throws at you most often, or chip away at your weak areas. **The best prep plan is the one that fits your timeline before test day.**

P1: Answer: B. The passage focuses on Maria's study of monarch butterfly migration and discusses possible influences on their routes, such as seasonal changes, wind patterns, and habitat destruction. These details support the broader idea of complex factors affecting migration, making B the correct answer.

P2: Answer: A. The passage highlights the growing popularity of urban gardens and outlines several benefits, including providing fresh produce, fostering community, and encouraging sustainability. These points all connect back to the main idea that urban gardening is on the rise and offers multiple advantages.

P3: Answer: C. The passage focuses on the coach's positive outlook despite a loss. Words like "pride," "growth," and "stepping stone" indicate an optimistic and supportive tone, making C the best choice.

P4: Answer: B. The author highlights dissatisfaction with city officials and warns about severe outcomes if action isn't taken. The language reflects both criticism and urgency, which aligns perfectly with B.

P5: Answer: A. The passage states that over 70% of visitors selected the paleontology tour, which strongly supports the conclusion that fossil exhibits are most popular.

P6: Answer: B. Mystery novels consistently outsold other genres for three years, so expanding that section logically matches customer interests.

PA1: B. The passage primarily highlights Wilson's methodical, scientific approach to ornithology, his influence on the field, and how his accuracy in depiction shaped modern scientific illustration. While Audubon is mentioned, it's only for contrast and not the main focus (eliminate A). There's no emphasis on poetry or journal writing beyond a brief reference (eliminate C). There's no rivalry explored (eliminate D).

PA2: B. The author praises Wilson's methodical approach and notes his influence on others, indicating a respectful tone.

PA3: B. 'Flamboyant' refers to someone who is flashy or showy, which contrasts with Wilson's understated approach.

PA4: C. The passage highlights Wilson's impact and contributions to the field of ornithology.

PA5: C. Bonaparte is mentioned as someone inspired by Wilson, demonstrating Wilson's lasting scientific influence.

PB1: B. The passage explains that urban planning developed in response to problems like overcrowding and sanitation.

PB2: C. The City Beautiful movement emphasized beautification and monumental grandeur.

PB3: D. Agricultural reform is not mentioned as a concern.

PB4: A. The passage states planners believed beautification would inspire civic pride and virtue.

PB5: B. The movement was criticized for focusing on looks rather than social improvements.

PC1: B. The passage clearly cites industrialization and population growth as reasons for planning reform.

PC2: C. The City Beautiful movement hoped beauty would inspire civic virtue and calm social unrest.

PC3: D. The passage mentions the displacement of marginalized communities as a critique.

PC4: B. Olmsted Jr. is described as formalizing urban planning as a profession.

PC5: B. The passage tracks a shift from unregulated to more balanced and inclusive planning.

PD1: B. The passage primarily outlines how Modernist movements emerged and what characterized them.

PD2: C. Kandinsky is cited as using color and form to evoke emotion, not represent the real world.

PD3: B. Fragmentation refers to the breaking apart of visual elements, as used in Cubism.

PD4: B. The Dadaists used found objects and presented them in provocative ways.

PD5: C. Modernism embraced new, often unconventional approaches and ideas.

PE1: B. Georgia O'Keeffe's impact on modernist painting and her artistic intent – The entire passage revolves around her distinct approach and legacy.

PE2: C. She rejected symbolic interpretations and clarified her intent – The passage states that she denied symbolic readings and emphasized visual appreciation.

PE3: C. a groundbreaking and influential figure – The author notes she was a pioneer and helped future female artists.

PE4: B. Hyperbole – 'Larger-than-life' is an exaggeration used to draw attention to the massive scale of her floral works.

PE5: C. She set a precedent for female artists to define their own narratives – The passage highlights her influence on later generations.

PF1: C. The passage centers on Amara's tryout and her internal and external experiences during that moment.

PF2: C. Amara finishes the tryout feeling proud that she gave her best effort, regardless of the result.

PF3: C. Her shoes symbolize her readiness and belief that preparation helps her succeed.

PF4: D. Tasha's intensity pushes Amara to step up and stay focused, showing internal motivation.

PF5: C. Coach Valdez's nod and raised eyebrows suggest he was impressed with Amara's smart pass.

PG1: B. The passage focuses on Jordan adapting to life in the city, making friends, and growing confident.

PG2: B. She is overwhelmed by trains, crowds, and even coffee shops, indicating confusion and anxiety

PG3: C. Zoe becomes Jordan's supportive friend and helps her feel more at home.

PG4: D. By spring, she knows her way around and doesn't feel like she's pretending anymore.

PG5: A. Pineville is described in contrast to the bustling, confusing city life.

PH1: C. The author repeatedly argues that preservation supports cultural continuity and reduces environmental waste

PH2: B. The phrase 'progress with perspective' implies blending innovation with respect for historical lessons.

PH3: D. The author mentions that reusing old structures reduces the need for demolition and new materials.

PH4: C. The author uses emotionally charged language to argue passionately for preservation.

PH5: A. This phrase highlights the aesthetic and artistic significance of historic architecture.

PI1: A. Both women succeeded in their fields despite the social and institutional limitations placed on them due to their gender.

PI2: C. The passage notes that Hayden left architecture, suggesting societal pressure and limited opportunities contributed to this outcome.

PI3: B. The passage explicitly states that Hurston's work was 'largely unrecognized in her lifetime.'

PI4: B. Each woman operated in male-dominated spaces and faced societal challenges related to their gender.

PI5: D. Though their fields differed, both Hayden and Hurston contended with societal limitations that shaped their careers.

1. B. The passage illustrates perseverance and personal growth even without recognition, not resentment or unfairness.
2. H. Miss Walker's words motivate the narrator to take the essay seriously.
3. C. His family neither actively discouraged nor strongly supported; they mostly allowed him to pursue it alone.
4. F. The phrase means he kept revising despite discouragement.
5. B. He was disappointed but outwardly composed (he clapped with others).
6. F. He realizes that perseverance itself is victory.
7. B. The description vividly sets the scene for the reader.
8. G. By the end, he has quiet acceptance and growth
9. A. The contrast between the polished essay and his ragged draft is emphasized.
10. J. The broader theme is perseverance and inner accomplishment rather than prizes.
11. C. The narrator reflects on exclusion but also on hope through education and clarity.
12. F. The stranger who ignored him revealed his growing awareness of exclusion.
13. B. The "veil" symbolizes separation from full participation in society.
14. G. He resists by using education as a weapon.
15. B. He treats books as tools for resistance, not casual pastimes.
16. H. "Second sight" means perceiving both reality and possibility.
17. A. The "curtain" imagery underscores separation and confinement.
18. J. By the end, he acknowledges the veil but frames it as a source of clarity.
19. A. The narrator recalls that in school, he could play games and share classroom activities equally with his peers, yet he also recognized a deeper sense of exclusion when treated differently by others (such as the visiting stranger). This contrast between outward participation and inner separation is a central theme of the passage.
20. F. Education as both refuge and resistance is the central theme.

High ACT reading skills, which encompass strong reading comprehension, critical analysis, and attention to detail, are valuable in numerous careers. While employers rarely ask for an ACT score directly, these abilities are essential for professions that involve complex reading, research, and communication.

10 jobs that require high ACT reading skills:

1. **Lawyer:** Lawyers must read and analyze vast amounts of legal documents, including statutes, case law, and contracts. Excellent reading comprehension is critical for building a case and presenting a legal argument.

2. **Editor:** Editors work for book publishers, magazines, and digital media companies, reviewing and revising written content. They need high-level reading skills to check for grammar, spelling, and factual errors, and to ensure clarity and style consistency.

3. **Researcher:** A researcher, whether in academia, market research, or scientific fields, must read reports, articles, and studies to gather information and draw conclusions. This requires synthesizing complex information and thinking critically about findings.

4. **Technical writer:** Technical writers create instruction manuals, how-to guides, and other documents that translate complex information into easily digestible content. They need superior comprehension skills to understand technical subjects and present them clearly to a non-expert audience.

5. Journalist: Journalists research and write news stories, features, and reports for various publications. They must be able to read and understand many sources of information, conduct interviews, and adhere to ethical standards for accuracy.

6. **Professor:** Professors in higher education must be avid readers to prepare their lesson plans and conduct research. They also read student work, including papers and projects, and grade them for understanding and clarity.

7. **Librarian:** Librarians do much more than organize books. They are information specialists who help people locate and evaluate research resources. They also manage library collections and promote literacy in their community.

8. **Grant writer:** Grant writers help organizations, typically non-profits, write proposals to secure funding. This involves extensive reading of grant applications to understand requirements and the effective writing of proposals to persuade funding sources.

9. **Archivist:** Archivists preserve and organize historical documents and other items for museums, libraries, and government institutions. This job is ideal for individuals who are fascinated by history and have a talent for interpreting and organizing old writings.

10. **Literary agent:** A literary agent represents authors to help them get published. This involves reading and evaluating numerous manuscripts to identify those that are likely to be successful in the market. They also manage contracts and communicate with publishers.

SCIENCE
H
Be
L
Li
L
Li
Be

Chapter 10: Science Stretch, warm-up, and drills

Science is optional on the ACT, and how much you invest in it depends on your strengths. If you're solid in science, put in the reps and keep building that muscle. If it's not your thing, your time is better spent sharpening English, Reading, and Math—the core of your score.

The good news? The ACT Science test isn't about memorizing biology terms or recalling chemistry formulas. It's mostly about **reading and reasoning**—charts, graphs, figures, maps, and labels. Your job is to locate information quickly and accurately.

Yes, you'll see some math, but if you're cranking out long calculations or wishing for a calculator, you've gone off track. The answers are almost always hidden in the visuals or the passage itself. Sometimes you'll need to extend a data trend, but that's just following the trail they've already laid out.

Let's kick off this Science stretch with something fun—a label. Who's hungry? What's your go-to snack? Today, we're breaking down a popular pick you'll spot on school campuses everywhere: **Takis.**

Nutrition Facts

About 25 servings per container
Serving size 1 oz (28g/about 12 pieces)

Amount per serving
Calories 150

	% Daily Value*
Total Fat 8g	10%
Saturated Fat 2.5g	13%
Trans Fat 0g	
Polyunsaturated Fat 3g	
Monounsaturated Fat 2.5g	
Cholesterol 0mg	0%
Sodium 420mg	18%
Total Carbohydrate 17g	6%
Dietary Fiber 1g	4%
Total Sugars 1g	
Includes 0g Added Sugars	0%
Protein 2g	
Vitamin D 0mcg	0%
Calcium 15mg	2%
Iron 0mg	0%
Potassium 54mg	2%

* The % Daily Value tells you how much a nutrient in a serving of food contributes to a daily diet. 2,000 calories a day is used for general nutrition advice.

1. About how many calories would you get from 36 pieces?
2. About how many grams are in the entire bag?
3. If your doctor said you're low on "Fe," would eating this bag help? Explain your answer.
4. What's the number of calories a day used for general nutrition advice?
5. How many calories are in the entire bag?

Mount Sano, a volcanic mountain on the Pacific Rim, has been studied by scientists dating back to the early 1900's. During that time period, scientists have studied eruption patterns and emission rates. The eruption dates are outlined in Figure 1 below. Using an optical correlation spectrometer and infrared analyzer, scientists were able to monitor the average Sulfur Dioxide (SO_2) and Carbon Dioxide (CO_2) emissions from Mount Sano both at the time of eruptions and during dormant periods in order to determine the effect on the Earth's climate (Figure 2). Scientists also studied the composition of volcanic gases and compared with other volcanoes in the region (Table 1).

Figure 1

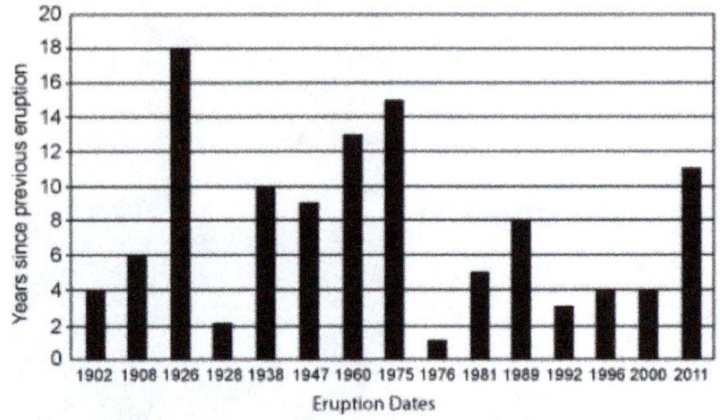

Figure 2

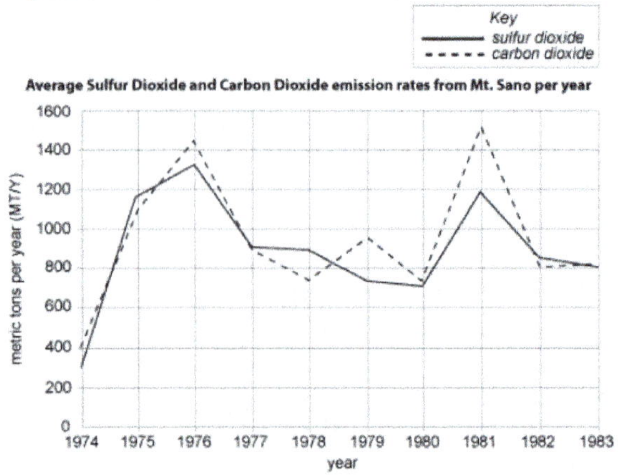

Table 1: Volcanic gas compositions, in volume percent concentrations for three different volcanoes

Volcano	Mount Sano	Mount Buxton	Mount Meyer
H_20	38.20%	76.30%	95.30%
CO_2	49.10%	10.80%	1.56%
SO_2	9.90%	8.10%	0.49%
H_2	0.80%	2.59%	0.71%
CO	1.21%	0.48%	0.01%
H_2S	0.69%	0.91%	0.24%
HCl	0.09%	0.82%	1.31%
HF	0.01%	0.00%	0.38%

36. How many volcanic eruptions took place at Mount Sano between 1980 and 2002?

A. 5
B. 6
C. 20
D. 24

37. Since 1902, how many dormancy periods of 10 years or longer were recorded for Mount Sano?

F. 2
G. 4
H. 5
J. 11

38. In Figure 2, what is the primary cause of the higher SO_2 and CO_2 levels in 1975, 1976, and 1981?

A. The temperature of the volcano was highest at these times
B. These were years where the volcano erupted
C. These were dormant periods for the volcano
D. Pollution caused by human activity caused the increase in SO_2 and CO_2 levels.

39. According to Table 1, which volcano produces the lowest levels of Sulfur Dioxide?

F. Mount Sano
G. Mount Buxton
H. Mount Meyer
J. Cannot be determined

40. In 1975, the CO_2 level of Mount Sano was measured at approximately 1,000 MT/Y. According to Figure 2 and Table 1, what was the approximate total amount in MT/Y of all gas emissions from Mount Sano in 1975, assuming that the same concentration percentages apply?

A. approximately 5,000 MT/Y
B. approximately 500 MT/Y
C. approximately 2,000 MT/Y
D. D. approximately 10,000 MT/Y

That was an example of a **Data Representation** passage. Questions are based on the graphs. Even if you wanted to read the intro, it wouldn't take long. For these, you can usually go straight to the question, and it will point you in the right direction. Example: "Based on figure 1" or "according to figure 2." Next, let's look at a **Conflicting Viewpoints** one. This is the opposite of data representation because it either doesn't contain an image/chart/figure at all, or if it does, the figure isn't useful. It's meant to waste your time and/or confuse you. Conflicting viewpoint questions involve two individuals with differing viewpoints/hypotheses. There will only be ONE (or sometimes none) of these questions on your test. Treat it like a compare-and-contrast reading passage. Slow down just enough to see where they agree, where they disagree, and how the questions frame those differences.

Artificial Intelligence (AI)

Artificial Intelligence (AI) refers to the development of machines and software that can perform tasks requiring human-like intelligence, such as decision-making, problem-solving, and language processing. While AI has already transformed industries ranging from healthcare to finance, researchers debate whether its continued growth will benefit or harm society in the long run. Two scientist give their professional opinions on AI.

Scientist 1

AI will become a positive force in society, provided that safeguards are in place. Like previous waves of technological change, AI may temporarily disrupt jobs, but it will also create entirely new industries and opportunities. For example, automation in manufacturing has historically eliminated repetitive labor but expanded demand for engineers, designers, and service providers.

Ethical concerns such as privacy and algorithmic bias can be managed through regulation and transparency. If governments require companies to disclose how data is collected and how AI systems are trained, unfair outcomes can be minimized. Far from being a threat, AI is a tool to increase productivity, accelerate medical research, and solve problems too complex for humans alone.

Scientist 2

AI represents a serious long-term risk. Unlike past technological revolutions, AI can automate not only physical labor but also cognitive tasks. This means that large numbers of jobs across multiple industries could vanish permanently. Unlike earlier shifts, there may not be enough new roles to absorb displaced workers.

Ethical risks are also more difficult to solve than most people think. AI systems learn from existing data, which often contains bias, meaning discrimination can be reinforced at massive scale. Privacy violations will only grow as AI processes ever-larger amounts of personal data. Left unchecked, AI could deepen inequality and concentrate power in the hands of a few corporations, threatening democratic values and human autonomy.

1. Based on Scientist 1's viewpoint, what role do regulations play in the development of AI?

A. They are unnecessary because AI will naturally correct itself.

B. They are essential to managing risks such as bias and privacy violations.

C. They prevent AI from reaching its full economic potential.

D. They should focus only on job creation, not on ethics.

2. Scientist 2 would most likely argue that the difference between AI and earlier technological revolutions is that AI:

F. automates both physical and cognitive tasks.

G. affects only highly skilled workers.

H. creates more jobs than it eliminates.

J. does not require large amounts of data.

3. Which of the following statements would both scientists most likely agree on?

A. AI has the potential to transform multiple industries.

B. AI will permanently eliminate more jobs than it creates.

C. Privacy concerns related to AI are overstated.

D. AI should be banned before widespread adoption.

4. According to Scientist 1, what is a likely outcome of automation?

F. Job losses with no new opportunities.

G. Repetitive jobs being replaced while new industries emerge.

H. AI reinforcing discrimination on a large scale.

J. Democratic values being threatened by concentration of power.

5. Which of the following best summarizes Scientist 2's concern about bias in AI?

A. Bias can be fully eliminated through transparent training methods.

B. Bias in AI is less harmful than bias in human decision-making.

C. AI bias is difficult to remove and could amplify existing discrimination.

D. Bias in AI occurs only when no regulations are in place.

6. Suppose that future studies show AI consistently improves fairness in hiring decisions compared to human managers. This finding would most weaken which scientist's position?

F. Scientist 1

G. Scientist 2

H. Both Scientist 1 and Scientist 2

J. Neither Scientist 1 nor Scientist 2

7. Which of the following best captures the central disagreement between the two scientists?

A. Whether AI should be used in manufacturing.

B. Whether AI represents a tool for solving problems or a threat to society.

C. Whether privacy issues related to AI are important.

D. Whether governments should regulate new technologies.

Finally, we've got **Research Summaries.** These appear more than any other Science question type. You'll see a mix of charts, tables, and graphs paired with short descriptions of experiments. A quick giveaway? Look for italicized words like *study* or *experiment*. That's your signal you're in Research Summary territory. Let's dive into one.

A student conducted two experiments to study the frictional force between surfaces. For each experiment, a block was pulled on a flat surface at a constant speed. The block was pulled with a spring scale, which was used to measure the force applied to the block, as shown in Diagram 1. The applied force measured on the spring scale is equal to the friction force between the block and the surface.

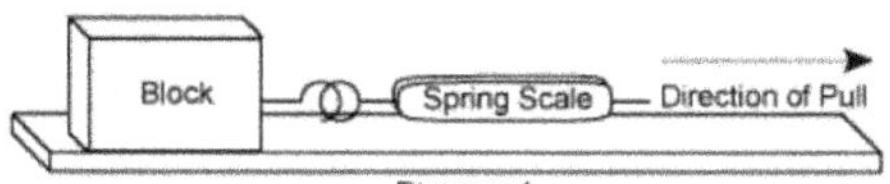

Diagram 1

The force of friction between a moving object and a surface can be described with the following equation. The coefficient of friction, μ, is an experimentally determined constant specific to each object and surface.

$$\text{Force} = (\mu)(\text{mass of block})(9.8)$$

Experiment 1

In experiment 1, wood blocks of different masses were pulled along sand paper. Each block had the same dimensions and surface area. Four different types of sand paper were tested. The force applied to the block was measured on the spring scale as the block was pulled at a constant speed. The results are shown in Figure 1.

Figure 1

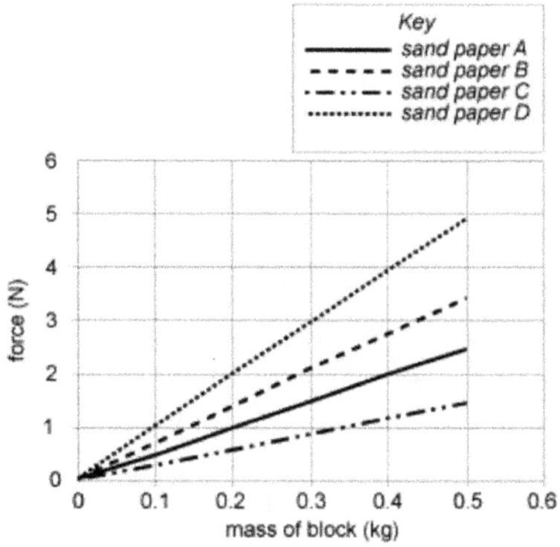

Experiment 2

In a second experiment, the student measured the force required to slide blocks of different materials across one type of sand paper. Each block had the same surface area in contact with the sand paper, and each block had a mass of 0.4 kg.

Table 1

Block Material	Force (N)
Wood	2.0
Rubber	4.2
Glass	0.73
Steel	0.95

20. In experiment 1, what force would be measured if a wood block with a mass of 0.6 kg was pulled across sand paper B?

A. 1.8 N
B. 2.9 N
C. 4.1 N
D. 5.9 N

21. Based on the results of experiment 1, which type of sand paper was used in experiment 2?

F. A
G. B
H. C
J. D

22. Based on both experiments, which sandpaper and block material combination would be expected to have the least friction of all the possible combinations?

A. Rubber and sand paper A
B. Wood and sand paper B
C. Glass and sand paper C
D. Steel and sand paper D

23. The student tested a fifth type of sand paper, Sand Paper E. The student measured a force of 4 N when using a 0.5 kg block. Based on the results in experiment 1, which of the following correctly lists the five types of sand paper in order of increasing friction force between the sand paper and the wood block?

F. A B C D E
G. B D E A C
H. D E B A C
J. C A B E D

24. According to experiment 1, if the mass of a block is doubled, the force of friction would _______.

A. Double
B. Triple
C. Halve
D. Stay the same

25. Based on experiment 1, what is the coefficient of friction for the wood block on sand paper D?

F. 0.1
G. 0.5
H. 1.0
J. 10

SCOUTING
REPORT
KNOW THE
PLAY.
MASTER THE GAME.
MOORE SUBSTANCE
TEST PREP

ACT Science Section: Smart Strategy Guide

1. Survey the landscape, then Question-Hunt

Skim (and ANNOTATE key things) the intro and passage, scan the charts/graphs, then go straight to the questions. Let the questions guide your focus. Use them as your roadmap to success.

2. Master the Graph Game

Pay attention to axes, units, trends, and legends. Understand how variables interact visually.

3. Use Keywords as GPS

Use keywords in the question to locate answers in graphs, charts, or tables.

4. Eliminate Emotion, Choose Logic

Don't guess based on what "sounds" right — rely only on what the data shows.

5. Watch for Comparison & Change

Spot differences between experiments or trends between graphs.

6. Don't Fear the Jargon

Skip confusing science terms unless the question specifically asks about them.

7. Know the Question Types

- Data Representation – Graph-based
- Research Summaries – Experimental setups
- Conflicting Viewpoints – Scientists' opinions

📝 Notes:

____

______

____

Chapter 11: Science Scrimmage

You've stretched, warmed up, and drilled. Now it's time to scrimmage in the Science arena. Remember—this isn't about things you've learned and had to memorize in your science classes. This is about speed, timing, focus, and smart plays with charts, graphs, and experiments.

In a meaningful practice, scrimmages are where prior preparation comes together—you run your plays under pressure and learn how to adjust on the fly. Same here. Every question you miss gives you film to study, and every question you nail builds confidence for game day.

You've seen the three types of questions you'll encounter on the ACT portion and had reps with them all. That said, we'll finish the day with one science set (you earned it): seven questions in seven minutes.

Avat Pharmaceuticals developed a GLP-1 agonist drug (Lowzempic) to help with weight loss. They did a clinical trial with 400 people in a controlled environment and collected the data for five months. Everyone who volunteered for the trial weighed 300-305 pounds and was given one of four options, unbeknownst to them. Group P was given a placebo pill and had to exercise for 30 minutes a day and maintain a healthy diet. Group P1 was given a placebo pill and didn't exercise or eat a monitored healthy diet. Group L was given a specific dose of Lowzempic and had to exercise 30 minutes a day and maintain a healthy diet. Group L1 was given a specific dose of Lowzempic and didn't have to exercise or eat a monitored healthy diet. There were 100 individuals in each group. Figure 1 has those results.

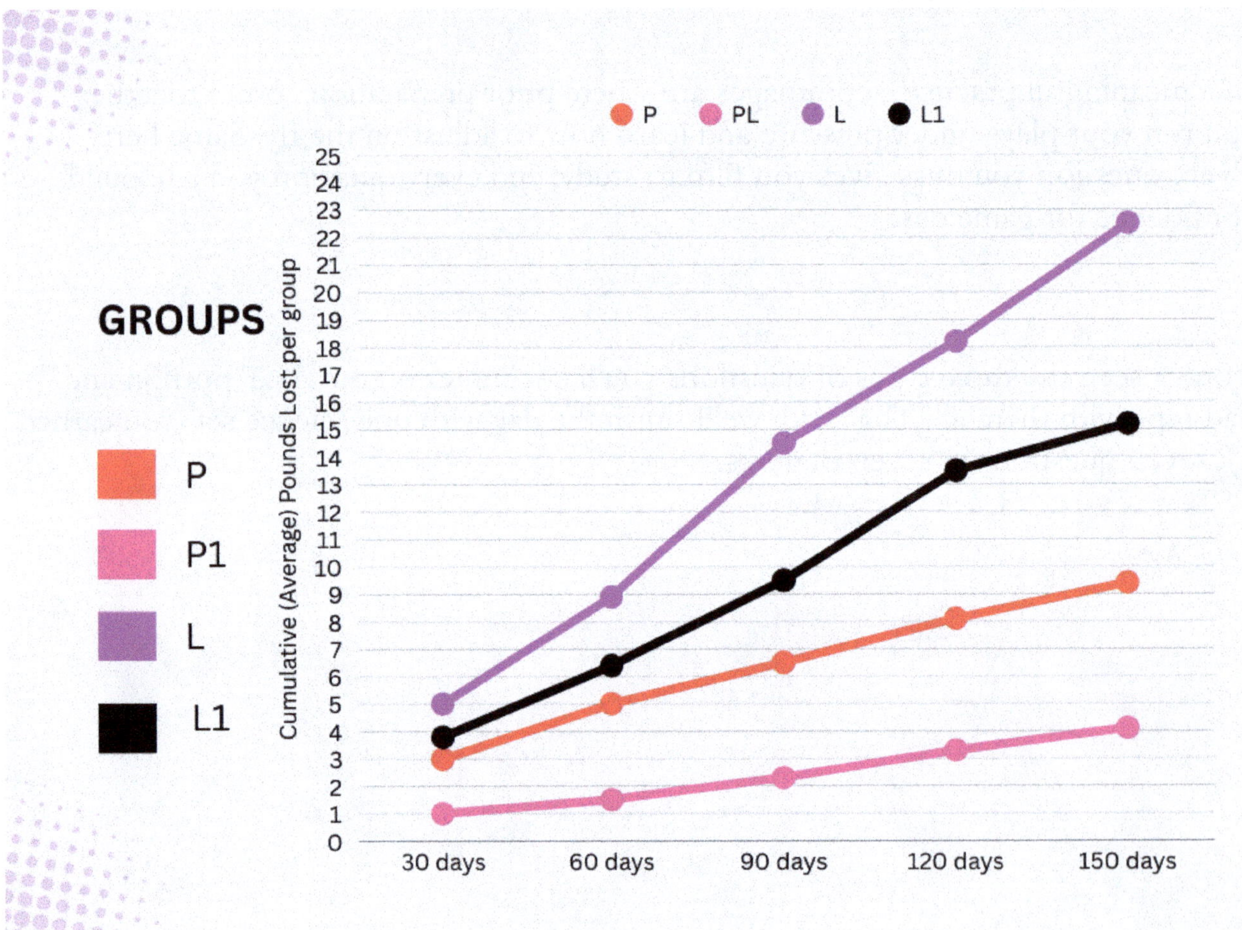

In addition to group comparisons, researchers also tracked outcomes across age ranges. Figure 2 shows the **average pounds lost after 150 days** in four age groups (20–29, 30–39, 40–49, 50–59) for participants who received Lowzempic. Results are displayed separately for Group L (drug + diet/exercise) and Group L1 (drug only)

Figure 2

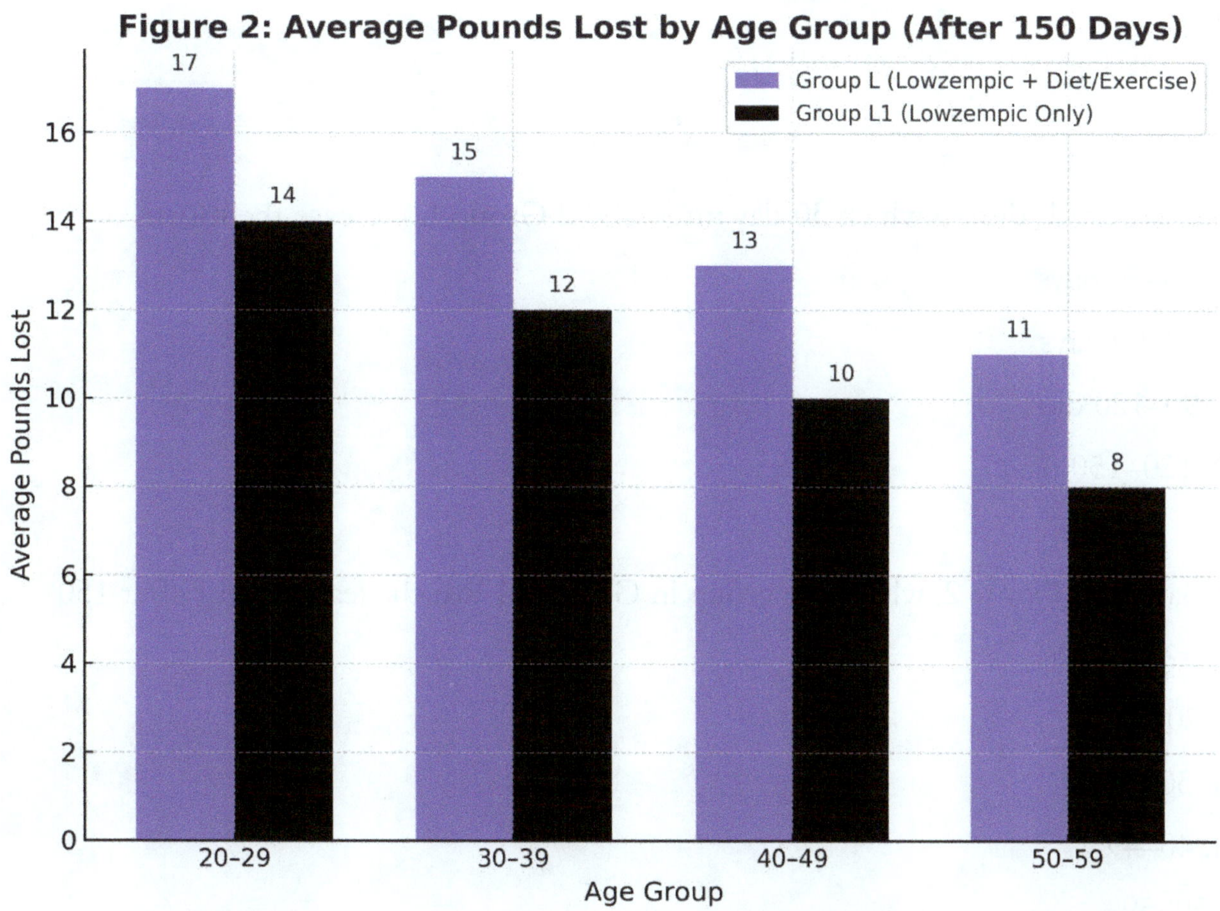

1. According to Figure 1, which group showed the greatest cumulative average weight loss at 150 days?

A. P

B. P1

C. L

D. L1

2. From 60 to 120 days, which group had the largest increase in cumulative average pounds lost (Figure 1)?

F. P

G. P1

H. L

J. L1

3. In Figure 1, during which 30-day interval did Group L increase the most?

A. 30–60 days

B. 60–90 days

C. 90–120 days

D. 120–150 days

4. Based on Figure 2, which age group in Group L1 lost the least weight after 150 days?

F. 20–29

G. 30–39

H. 40–49

J. 50–59

5. Which statement is best supported by Figure 1?

A. Group P1 surpassed Group P by 120 days.

B. Group L1 exceeded Group P at every time point.

C. Group L plateaued after 90 days.

D. Group P increased faster than Group L from 90–150 days.

6. Suppose a new Group L2 (double-dose Lowzempic, no lifestyle changes) produced results 20% higher than Group L1 at each time point. Based on Figure 1, the expected cumulative average loss for L2 at 150 days would be closest to:

F. 14 lb

G. 16 lb

H. 18 lb

J. 22 lb

7. Using Figure 2, how many age-group × treatment combinations (out of 8) met or exceeded 12 lbs. of average loss after 150 days?

A. 3

B. 4

C. 5

D. 6

CONGRATULATIONS!!! You made it. You've stretched, warmed up, drilled, and scrimmaged through every part of the ACT. That's the full playbook. From English to Math, Reading to Science, you've put in the reps. Now it's about execution on game day.

Here's your final prep:

- **Rest matters.** Get a full night's sleep before test day. Don't cram until 2 a.m.—fatigue kills focus faster than a tough question.

- **Fuel up.** Eat a balanced breakfast—protein, fruit, water. Skip the sugar spikes and the extra caffeine that'll leave you crashing halfway through the test. During your break, stretch, get some water, use the bathroom, or have a healthy snack. Avoid sugar-filled treats.

- **Pack smart.** Bring your admission ticket, photo ID, #2 pencils, an approved watch, and an approved calculator that you're already familiar with.

- **Breathe.** When the pressure hits, remember your drills. Pace yourself, trust your training, and one question at a time, you'll work the plan.

- Do **NOT** check your phone during the break. If something great is happening, you won't want to put it down. If something tragic has transpired, you'll be hard-pressed to refocus. Either way, you can catch up after the test.

The ACT doesn't measure your worth—it's just one game. But if you play it smart, rested, and confident, you'll walk off knowing you gave it your best and potentially earn a whole lot of cash to use for college. And that's a win, WIN.

Next Steps: How to Keep Preparing

You've completed the scrimmages—great work. But your prep doesn't stop here. There are different ways to continue sharpening your ACT skills. Below, you'll find the Focus Areas for Science. It highlights which topics tend to show up the most on the test. Pair that with one of three strategies to guide your study plan:

1. **Double Down on Your Strengths**
   Keep practicing what you're already strong at. This locks in easy points and helps you move quickly on test day, freeing time for harder questions.

2. **Prioritize High-Frequency Topics**
   Focus on what shows up most often. Even if you're not perfect, being solid here guarantees a higher return for your effort.

3. **Tackle Your Weaknesses**
   If there's an area that consistently slows you down, spend time there. Improving even a little can boost your confidence and raise your score.

No single approach is "best"—the right choice depends on what you think is best (you know yourself better than anyone else) and how much time you have before test day.

Science Focus Areas

The top tested areas in the ACT Science section are Interpretation of Data (40–50%), Evaluation of Models, Inferences, and Experimental Results (25–35%), and Scientific Investigation (20–30%), with a focus on scientific reasoning and problem-solving skills rather than knowledge of specific scientific disciplines. The test covers topics in biology, chemistry, Earth/space science, and physics, but doesn't require advanced knowledge.

Tested Areas

Interpretation of Data (40–50%)

: This is the most heavily weighted area, involving reading and understanding scientific data presented in charts, graphs, and tables to draw conclusions and identify relationships.

Evaluation of Models, Inferences, and Experimental Results (25–35%)

This section focuses on assessing scientific methods and reasoning by examining experimental setups, evaluating experimental results, and drawing logical inferences from the data.

Scientific Investigation (20–30%)

This area tests your ability to understand the design of scientific studies, predict the outcomes of experiments, and make judgments about the quality and limitations of scientific findings.

Key Aspects to Focus On

Scientific Reasoning over Specific Knowledge: The test emphasizes your ability to apply scientific thinking to new information rather than recalling facts from a particular science course.

Data and Graph Comprehension: A strong understanding of how to interpret and analyze graphs and data sets is crucial, as this is a significant portion of the test.

Understanding Experimental Design: You need to be able to understand how experiments are set up and how to evaluate their results and the conclusions drawn from them.

From here, you can decide what to do: strengthen your strong spots, focus on what the test throws at you most often, or chip away at your weak areas. **The best prep plan is the one that fits your timeline before test day.**

1. 450

2. 700

3. "Fe" represents iron on the periodic table, and it's 0mg of iron in the bag.

4. 2,000

5. 3,750

36. From 1980 to 2000, five eruption dates are on Figure 1.

37. Figure 1 (y axis) shows years since previous eruption, and 5 of them show 10 or higher.

38. According to Figures 1 and 2, the eruption years coincide with the gas spikes.

39. Table 1 shows that 0.49% has the lowest SO_2.

40. According to Table 1, CO2 makes up 49.1 % (almost 50%) of the gas emissions from the volcano. If 50% is 1,000 MT/Y, then the remaining 50% would double that.

1. B. Scientist 1 states that ethical concerns can be managed through regulation and transparency.

2. F. Scientist 2 emphasizes that AI is unique because it can replace not just manual labor but also cognitive work.

3. A. Both scientists agree AI will have major impacts. They disagree on whether these impacts will be mostly positive or negative.

4. G. Scientist 1 compares AI to past technologies, arguing that while some jobs are lost, new roles and industries will appear.

5. C. Scientist 2 argues that AI learns from biased data, which can reinforce discrimination on a large scale.

6. G. If AI studies showed improved fairness in hiring, this would directly weaken Scientist 2's argument that AI bias is difficult to manage and dangerous.

7. B. Scientist 1 views AI as a tool to solve problems and boost productivity. Scientist 2 views it as a long-term threat to jobs, fairness, and autonomy.

20. C. At 0.5kg, it's already at 3.5N. At the increasing rate, it's more likely it'll be at 4.1N, not 5.9N.

21. F. 0.4kg was used for experiment 2, and the measured force was 2.0N. Based on Figure 1, that matches sandpaper D.

22. C. Both glass and sandpaper C cause the least amount of friction when compared to others.

23. J. The lowest friction is C, so no matter what is added, it must start with C. Use process of elimination.

24. A. In the given equation for force, if one thing doubles, the total will double since it's being multiplied.

25. H. 0.4kg wood and sandpaper D = 4.0N. round 9.8 to 10 for easier math, and multiply mass times 10 to get 4.0, so the coefficient must be 1 to still have approximately 4.0N.

1. C. In Figure 1, the purple L line ends highest (~23 lb) at 150 days.

2. H. Between 60 and 120 days: L jumps from ~9 lb to ~18 lb (≈ +9), the largest increase. L1 rises ~+7, P ~+3, P1 ~+2.

3. B.

4. J. In Figure 2, L1 bars are 14, 12, 10, and 8; the smallest is 50–59.

5. B. Figure 1 shows the black L1 line above the red P line at 30, 60, 90, 120, and 150 days. The other statements contradict the graph.

6. H. At 150 days, L1 is about 15 lb. A 20% increase gives 1.20 × 15 ≈ 18 lb. (break 15 into five parts to get 20% for each part. Each part is 3.)

7. C. Group L ≥12 lbs. for 20–29, 30–39, 40–49 (but not 50–59) → 3. Group L1 ≥ 12 lbs. for 20–29 and 30–39 (exactly 12) → 2. Total = 5.

The ACT Science section is less about memorized facts and more about critical thinking, data analysis, and scientific reasoning. Therefore, the jobs that "require" an ACT Science score are those within competitive college programs that use the score as an indicator of a student's aptitude for STEM (science, technology, engineering, and mathematics) fields. A high score can help you get into a challenging undergraduate program in a specific field, such as:

- **Engineering**: Chemical, Biomedical, Aerospace.
- **Computer and Information Science**: Data Science, AI Research.
- **Medical Sciences**: Pre-med programs, Medical Scientist.
- **Physical Sciences**: Physics, Astronomy.
- **Environmental Science**: Geology, Hydrology.

10 jobs that demonstrate the skills measured by the ACT Science section, with many ranking among the highest-paying science jobs:

1. **Medical Scientist:** Medical scientists research diseases, develop and test treatments, and produce medical devices. Success in this field requires the ability to interpret complex data from clinical trials and experimental designs, a key skill tested on the ACT Science section.

2. **Physicist or Astronomer:** These scientists study the universe and its fundamental principles, using mathematical theories and high-tech equipment to analyze data. Their work demands strong skills in data representation and concluding scientific findings, skills that the ACT is designed to measure.

3. **Artificial Intelligence (AI) Research Scientist:** An AI research scientist develops machine learning and AI technologies, analyzing large datasets to train AI models. This field is directly reliant on the kind of data interpretation and critical reasoning emphasized by the ACT Science section.

4. **Geneticist or Genetic Counselor:** Geneticists study genes, genetic variation, and heredity, while counselors help interpret genetic data for patients. Both roles require strong analytical skills to work with complex genetic data, which align with the research summaries portion of the ACT Science test.

5. **Chemical Engineer:** Chemical engineers use principles of physics, chemistry, and engineering to design and develop manufacturing processes for various products. Their work involves the practical application of scientific principles, like the experimental reasoning required for the ACT Science section.

6. **Environmental Scientist:** Environmental scientists conduct research to identify and solve environmental and health problems. They collect, analyze, and interpret various forms of data (air, water, soil), which directly reflects the scientific skills tested by the ACT.

7. **Biomedical Engineer:** Combining engineering principles with medical science, biomedical engineers develop medical equipment like artificial organs and diagnostic tools. Their work involves interpreting and applying scientific data to practical problems, a direct parallel to the ACT Science section's focus on applied reasoning.

8. **Forensic Science Technician:** Forensic technicians collect and analyze evidence from crime scenes using chemical, biological, and physical sciences. The methodical and data-driven nature of this work aligns with the scientific investigation and analytical skills tested on the ACT.

9. **Pharmacist:** Pharmacists dispense medication and advise patients, collaborating with other medical professionals to ensure proper drug use. The role requires a strong scientific understanding of how drugs affect the body and the ability to interpret clinical data.

10. **Research and Development (R&D) Engineer:** An R&D engineer is responsible for designing new products and improving existing ones. The position requires using scientific methodology to conduct research and tests, a core skill for a high ACT Science score.

WRITING

Chapter 12: The ACT Writing Test

The ACT Writing Test is an optional section, but many colleges either recommend or require it, so it's important to check the admissions policies of the schools you're considering. Unlike the other sections of the ACT, which are multiple-choice, the Writing Test consists of one essay task. You'll be given a prompt that presents an issue and three different perspectives on that issue. Your job is to:

1. Clearly state your own perspective on the issue.
2. Analyze the relationship between your perspective and at least one of the provided perspectives.
3. Develop a well-organized essay with specific examples, logical reasoning, and strong support for your argument.

You'll have 40 minutes to plan, draft, and revise your essay. Two trained readers score the essay on four key domains: Ideas & Analysis, Development & Support, Organization, and Language Use & Conventions. Each domain receives a score from 2 to 12, and the two readers' scores are combined into your final Writing score.

How to Prepare for the Writing Test

- Practice timed writing: Get comfortable planning and drafting an essay in 40 minutes. Spend a few minutes outlining before you begin writing.
- Read sample prompts and essays: ACT.org provides examples of past prompts and scored essays—reviewing these helps you understand what graders are looking for.
- Strengthen your thesis statements: A clear thesis that takes a position is the backbone of a strong essay.

- Use specific evidence: Draw on history, literature, current events, or personal experiences to support your points.
- Practice transitions and organization: Make sure each paragraph flows logically into the next.
- Polish grammar and style: Clear, concise sentences with varied word choice make your essay easier to read and score higher.

Tip for test day: Don't panic if your first idea isn't perfect. Pick a perspective, commit to it, and focus on building a strong, logical argument. The graders are looking for clarity and depth of thought—not perfection.

Notes & Work Space

The top section of the page is lined for jotting down key reminders, formulas, or strategies in writing
The lower blank section is open for sketches, diagrams, outlines, math work, or quick problem-solving.

Notes & Work Space

The top section of the page is lined for jotting down key reminders, formulas, or strategies in writing
The lower blank section is open for sketches, diagrams, outlines, math work, or quick problem-solving.

Notes & Work Space

The top section of the page is lined for jotting down key reminders, formulas, or strategies in writing
The lower blank section is open for sketches, diagrams, outlines, math work, or quick problem-solving.

Notes & Work Space

The top section of the page is lined for jotting down key reminders, formulas, or strategies in writing
The lower blank section is open for sketches, diagrams, outlines, math work, or quick problem-solving.

Disclaimer

This book is an independent resource created to help students prepare for the ACT® test. It is not affiliated with or endorsed by ACT, Inc. ACT® is a registered trademark of ACT, Inc.

It includes two science passages from © *Sophia Learning, LLC.* (Science questions 36-40 and 20-25) Moore Substance Publishing does not own the rights to those questions.

These materials are provided strictly for educational and practice purposes.

Copyright Notice

www.ingramcontent.com/pod-product-compliance
Lightning Source LLC
Chambersburg PA
CBHW061214240926
51846CB00006B/1221
9781963933062